Routledge Revivals

Bad News

It is a commonly held belief that television news in Britain, on whatever channel, is more objective, more trustworthy, more neutral than press reporting. The illusion is exploded in this controversial study by the Glasgow University Media Group, originally published in 1976.

The authors undertook an exhaustive monitoring of all television broadcasts over 6 months, from January to June 1975, with particular focus upon industrial news broadcasts, the TUC, strikes and industrial action, business and economic affairs.

Their analysis showed how television news favours certain individuals by giving them more time and status. But their findings did not merely deny the neutrality of the news, they gave a new insight into the picture of industrial society that TV news constructs.

'The book deserves close study and establishes the fact that a value-free, "neutral" and exhaustively informative news is a myth'
- *Times Educational Supplement*

Bad News

Volume 1

Glasgow University Media Group

First published in 1976
by Routledge & Kegan Paul Ltd

This edition first published in 2010 by Routledge
2 Park Square, Milton Park, Abingdon, Oxon, OX14 4RN

Simultaneously published in the USA and Canada
by Routledge
270 Madison Avenue, New York, NY 10016

Routledge is an imprint of the Taylor & Francis Group, an informa business

© 1976 Glasgow University Media Group

All rights reserved. No part of this book may be reprinted or reproduced or utilised in any form or by any electronic, mechanical, or other means, now known or hereafter invented, including photocopying and recording, or in any information storage or retrieval system, without permission in writing from the publishers.

Publisher's Note

The publisher has gone to great lengths to ensure the quality of this reprint but points out that some imperfections in the original copies may be apparent.

Disclaimer

The publisher has made every effort to trace copyright holders and welcomes correspondence from those they have been unable to contact.

ISBN 13: 978-0-415-57268-2 (set)
ISBN 13: 978-0-415-56376-5 (hbk)
ISBN 13: 978-0-415-56787-9 (pbk)
ISBN 13: 978-0-203-09263-7 (ebk)

ISBN 10: 0-415-57268-1 (set)
ISBN 10: 0-415-56376-3 (hbk)
ISBN 10: 0-415-56787-4 (pbk)
ISBN 10: 0-203-09263-5 (ebk)

BAD NEWS

The authors are
Peter Beharrell
Howard Davis
John Eldridge
John Hewitt
Jean Oddie
Greg Philo
Paul Walton
Brian Winston

Previous works by the research team

Professor John Eldridge *Co-investigator*
Industrial Disputes, Routledge & Kegan Paul, 1972
Max Weber: The Interpretation of Social Reality
Michael Joseph, 1971/Nelson 1972
A Sociology of Organisations (with A. Crombie),
Allen & Unwin, 1975
Editorial Board of *Sociology and Industrial Relations Journal*

Paul Walton *Senior Lecturer in Sociology Co-Investigator*
Situating Marx (with S. Hall), Chaucer Press, 1971
From Alienation to Surplus Value (with A. Gamble)
Sheed & Ward, 1972/PB, 1976
The New Criminology (with J. Young and I. Taylor)
Routledge & Kegan Paul, 1973
Critical Criminology (with J. Young and I. Taylor)
Routledge & Kegan Paul, 1974
Capitalism in Crisis (with A. Gamble)
Macmillan, 1976
Editor *Theory and Society, Crime and Social Justice,*
Macmillan Library of Critical Studies

Brian Winston *Research Director*
Dangling Conversations – The Image of the Media
David Poynter, 1973
Dangling Conversations – Hardwear/Softwear
David Poynter, 1974

Volume 1

Glasgow University Media Group

Foreword by Dr Richard Hoggart

Routledge & Kegan Paul
London, Boston and Henley

First published in 1976
by Routledge & Kegan Paul Ltd
39 Store Street,
London WC1E 7DD,
Broadway House,
Newtown Road,
Henley-on-Thames,
Oxon RG9 1EN and
9 Park Street
Boston, Mass. 02108, USA
Reprinted in 1980
Reprinted and first published
as a paperback in 1981
Set in Monotype Times by Kelly Selwyn & Co,
Melksham, Wilts
and printed in Great Britain by
T. J. Press (Padstow) Ltd
© Glasgow University Media Group 1976
No part of this book may be reproduced in
any form without permission from the
publisher, except for the quotation of brief
passages in criticism

ISBN 0 7100 8489 7 (C)
ISBN 0 7100 0792 2 (P)

In Memory of Harvey Sacks

CONTENTS

FOREWORD

Dr Richard Hoggart,
Warden of Goldsmith's College,
University of London

The Glasgow Media Group have produced here one of the few detailed studies so far made of the way British television shapes what it calls 'the news'. Even before publication it has had a hostile or at best a suspicious reception from many people involved in that process. This was partly because the group issued a brief report on its findings to the Annan Committee on Broadcasting, a report which – because it had to be put together quickly – could not fully give all the evidence in support of what it said. There is also a more substantial reason for the hostility. None of us like our professional practices to be scrutinised by outsiders; and television newsmen must be near the top of the league in this kind of defensiveness. Sometimes I think the strength of their defensiveness is in direct proportion to their refusal to take a good look at just what they are doing each day. One gets the impression of a trade which has hardly ever thought out its own basic premises but continues, come hell or high water, to rest its case on a few unexamined assertions.

Such as the assertion that their news presentation is 'objective', a mirror of the reality outside, that they are merely neutral channels for presenting 'the facts', the nuggets of each day's hard news. This is so inadequate as an explanation of the complex process in which they are actually involved that you wonder how people in the news business, or their bosses at the top of the major instruments of broadcasting, can go on making it.

Of course, what they call 'the news' is biased; or, if that seems too loaded a word, artificially shaped. It is the result each day of a process of selection so speedy and habitual as to seem almost instinctive. There is simply too much possible material; there have to be filters, devices to select what shall be shown, in what order, at what length and with what stresses. Those devices have to be, as it were, in the blood; there isn't time to have a daily conference on first principles (a quarterly one would be an improvement, though).

'The news' selects itself by four main filtering processes. First, by simple constraints germane to the medium or fortuitously of the moment: constraints of available time or available resources, or of geography and so on. Second, 'the news' is decided by a tradition of 'news values' which television has largely taken over from the more popular end of the press and holds on to with little apparent will to think through their relevance, or irrelevance, to television's own situation (it wouldn't be a bad thing if the press itself had a closer look at some of its own assumptions too). Third, there are what are known as specifically 'television values' or 'television material'. This item is recognised as good visually, so it rates a place before that; or this subject will be approached thus because that's the angle which makes 'good television'.

The fourth and most important filter – since it partly contains the others – is the cultural air we breathe, the whole ideological atmosphere of our society, which tells us that some things can be said and others had best not be said. It is that whole and almost unconscious pressure towards implicitly affirming the status quo, towards confirming 'the ordinary man' in his existing attitudes, towards discouraging refusals to conform, that atmosphere which comes off the morning radio news-and-chat programmes as much as from the whole pattern of reader-visual background-and-words which is the context of the television news.

By the concurrent and almost instantaneous application of these four types of filter, television gives us what its practitioners call 'the objective news' but what is in reality a heavily-selected interpretation of events, one which structures reality for us, which shapes and frames a world for us to inhabit and accept as real and legitimated, one which sets the agenda within which – except by a positive effort at remaking – we are led to discuss the terms of our lives.

This is plainly a more interesting, promising and credible way of starting to think about the news process than the present shibboleths. The sooner the newsmen and their superiors accept this starting-point, the better for them, us and 'the news'. In this they can after all learn something from outsiders, even if outsiders don't know all the jargon of their trade, even if the outsiders are those always suspect academics or, worse, social scientist academics.

So this study sets out to reveal what the authors call 'some of the rules and structures governing the codings'; that is, some of the habitual ways in which there is chosen what will be called 'the news', from day to day. In recommending the book, in particular to anyone who is concerned with the making of television news, I am not suggesting that they should accept it uncritically. They should scrutinise it in the closest detail and be severe on any error, distortion, omission, bias they may find. But they should try to do so dispassionately, not in a mood of defensiveness so rancorous that they are unable to see the main drift of the argument, nor with an implicit urge to let niggles about this or that statistical point justify their resolute refusal to face the general case. The time is overdue for them to look steadily at that case.

On the other hand, those who make this kind of study have to steer clear of errors which are apparently just as tempting to many of them as is instant defensiveness to the producers. At its simple worst, this comes out as what one might call 'low conspiracy theory'. Here the fourth filter, that provided by the whole context and atmosphere of society itself, is seen as operating in a very direct way. It is assumed that orders are given that *this* shall be shown and *this* not, that telephone calls from high places decide what stresses there shall be and so on. There *are* sometimes pressures of this kind; it would be jejune to deny that they exist at all. But they are neither as frequent nor as important as some romantics would like to think.

The 'high conspiracy theory' takes you further. By this, you push aside the 'direct order from Downing Street' idea in favour of a much less direct approach. So you argue that the agenda is very tightly framed, in its inclusions, omissions and stresses, not by direct orders but by a number of more hidden forces, by – first – the process of recruitment to the profession itself (so that real dissidents to the prevailing concept of 'news' don't get in or don't stay in), by the unspoken but nevertheless firm transmission of the knowledge that *this* is how you tackle, say, questions about race, or strikes in the motor industry, or the troubles in Northern Ireland; that, where the media are so much *implicitly* controlled in their agenda-setting functions, direct pressures are rarely needed. Again, this is a line which can be very usefully followed a long way. But it does not take you the whole way, by any means. It is usually buttressed by some such statement as that in capitalist societies the powers that be control not only the means of production and distribution of goods, but also those for the production and distribution of ideas and comments; that therefore the agenda they set for 'the news' is one which almost wholly reinforces the *status quo.*

There are a lot of inadequacies showing by the time we have reached this point. First, that closed and reinforcing tendency, which certainly exists, is not a feature of capitalist societies alone or most strikingly. It is a feature of all large corporate states. Which is why the control of the media, the decisions as to what constitutes 'the news', are very much tighter in communist states than in commercial democracies. There, both 'low conspiracy' and 'high conspiracy' practices lock together neatly and very firmly. So those people, and I am not referring to the Glasgow Group of course, who begin to discuss news-structuring in a society such as Britain with a preamble like this: 'It is a feature of capitalist society that . . .' have got their eyes off the object, are being led more by their own political views than by the reality of broadcasting practices internationally.

But the basic inadequacy, even of the most subtle forms of 'high conspiracy theory', is that, useful and revealing though they can be, they miss the complexity of the matter if you hang on to them too long. For in commercial democracies such as Britain the agenda is not *wholly* structured and the deviant items are not simply 'permitted variations', the repressive tolerances of an authority which is nevertheless wholly in control in the background. Something sometimes escapes, precisely because the controls, explicit or implicit, are not complete, because the claims for objectivity and neutrality by the broadcasters – though often made too smugly and blandly – do also have behind them, in some people, a belief that the *effort* at objectivity and neutrality is important beyond all outside pressures. Some men act on principle, some out of whim, some in defiance, some from conviction; the process is not by any means wholly contained. Or where a bias does show it is not always ideologically motivated, but may be due to simplified cultural assumptions. I do not believe that the early stress on the health hazard in the Glasgow dustcart drivers' strike was inspired by a slightly sinister desire to put the strikers in a bad light; it was much more likely – especially because of the opportunities it gave for showing rats moving among the rubbish – to have been inspired by the trivial notion that this was a 'good news angle', 'good visual stuff', the sort of flesh-creeping, sensational material which 'they' out there are supposed to be tickled by. Similarly, I do not believe that the extraordinary preoccupation of television news with strikes and strikers in the motor industry – which sometimes makes you feel as though this whole island will soon sink bankrupt beneath the waves if 'a handful of men at Coventry' don't stop misbehaving and cutting off car production – I don't believe this preoccupation can be most accurately or illuminatingly explained in ideological terms (though it has those elements). Much more powerful is a mythical-cultural impulse, the extraordinary force of the motor car as a symbol in our kind of society, which creates an interest much greater than any that can be generated by pit-heads or the postal service.

At this point, one is beginning to define the four filters and their interconnections much more effectively. This is where the study of 'what makes news' and 'how the news is made' becomes most interesting and complex, both for students from outside and for the newsmen themselves. This is where it should start, rather than in a slanging match – with the newsmen well dug in from the beginning behind their barricades of old assumptions. We need, first, much closer and more varied analysis of the process by which 'the agenda' is set up, of what really happens, deep down and inside, and of the degree to which it is or isn't preordained. From that we can move to what also and no less needs doing – to analysing what really are the relations, direct or indirect, explicit or implicit, between the television people and the authorities – government, business, the trades unions. Those relations are tough and tender, strong and weak, and extraordinarily complicated; and they do not all run in the same direction. If we did work in depth on these two fronts we would at last be getting somewhere. The Glasgow Media Group would claim no more – and they have a right to claim it – than that they are contributing, at a relatively early stage, to that sort of understanding.

Richard Hoggart

ACKNOWLEDGMENTS

We would like to thank the following people for their encouragement, advice and help: Christine Douglas, Margaret Hall, Pru Larsen and Ann Taggart of the University of Glasgow; The Staff of the University Television Service and the University Administration for financial help and general support; the Social Science Research Council for financing the Project, and anonimity; the staff of the BBC's Audience Research Department: the staffs of the BBC's newsrooms in London and Glasgow, ITN and Scottish Television Limited; Dave Bridges; Tom Burns; Rex Cathcart; Umberto Eco; Robin Fox; David Frisby; Andrew Gamble; Erving Goffman; Al Gouldner; John Gray; Toni Griffiths; Stuart Hall; Caroline Heller; Richard Hoggart; Elihu Katz; Roy Lockett; Des McNulty; Graham Murdock; Brian Murphy; Henrietta Resler; Franco Rositi; Ian Roxborough; Alan Sapper; Anthony Smith; Ian Taylor; Laurie Taylor; John Westergaard; Steve White; Philip Whitehead; Colin Young; Jock Young.

We have also benefited from discussions and seminars at the following universities and organisations: the Institute of Education, the University of London; the Department of Sociology and the Centre for Mass Communication Research, the University of Leicester; the Institute for Workers Control, the University of Sheffield; the Industrial Relations Unit, the University of Nottingham; the Centre for Television Research, the University of Leeds; Department of Communications, Glasgow College of Technology; the Science and Technology Groups, the University of Sussex; Broadcasting Symposia, the University of Manchester; the National Deviancy Symposium, the University of Cardiff; the Criminology Unit, the University of Sheffield; British Sociological Association, Industrial Sociology and Mass Media Study Groups; the Academic

Section of the Prix Italia, 1975, at Florence; the Press and Publicity Departments of National and Regional Conferences of NALGO; interested members of the General Council of the TUC; the Sociology Departments of the Universities of Edinburgh and Glasgow; and, above all, the Centre for Contemporary Cultural Studies, the University of Birmingham.

Special mention must be made of Stella McGarrity for timely help; Yusuf Ahmad for adding a local dimension to the study; and Alison McNaughton who freely and unstintingly gave of her time and energy during the first year of the project and thereby contributed enormously to the work.

The Glasgow Media Group
The University
Glasgow

I REVIEWING THE NEWS

Reluctance to display its codes is a mark of bourgeois society and the mass culture which has developed from it.

R. Barthes

Contrary to the claims, conventions, and culture of television journalism, the news is not a neutral product. For television news is a cultural artifact; it is a sequence of socially manufactured messages, which carry many of the culturally dominant assumptions of our society. From the accents of the newscasters to the vocabulary of camera angles; from who gets on and what questions they are asked, via selection of stories to presentation of bulletins, the news is a highly mediated product.

This study is an attempt to unpack the coding of television news. It aims to reveal the structures of the cultural framework which underpins the production of apparently neutral news. At some common-sense level television news has to appear neutral or its credibility would evaporate.

There is the agreement widely shared in our society that television news is more objective than the press. Indeed most people still firmly believe that it is intrinsically more trustworthy. Historically, it is argued, the press is partisan, whilst broadcasting is more neutral. It is this illusion, this 'utopia of neutralism', that many of our findings deny. So thoroughly convinced are most media professionals and the public that television news is trustworthy that the basis of such beliefs must lie in the national culture. The public seem prepared to credit the national television organisations with a neutrality they would deny to others.

A BBC survey conducted in 1962 demonstrated that 58 per cent of the population use television as their main source of news, as against the 33 per cent who principally rely upon newspapers. Most significantly the survey showed that 68 per cent of the population interviewed believed that television news was the most trustworthy news medium, whilst only 6 per cent said this of the press.[1]

One perceptive critic, Trevor Pateman, has noted that whilst British audiences seem prepared to believe that foreign news services are faulty, biased, or distorted, the same judgment is not accorded our own news services. He suggests that this privileged assumption that our news service is best and is reliable, rests upon an unstated nationalism which historically is only made credible by the BBC's role during the Second World War.[2]

In the post-war period television has come into unchallenged dominance as the prime medium of entertainment and news communication. Such news is bound by law, and convention, to be balanced, impartial, unbiased and neutral. The news is produced on a day-to-day basis by a professional media élite who whilst doing their best, embody in their routine practices ideological assumptions which reinforce certain stratified cultural perceptions of society and how it should, ought, and does, work.

Much of the debate about the ideology of television production in general and news in particular would be unimportant were it not for the fact that television is now the front-runner medium eclipsing everything else bar face-to-face communication. In the UK most families spend four to five hours per day watching television, whilst on any given day around 60 per cent of the adult population hear or see at least one broadcast news bulletin. Moreover, until recent shifts in public attitudes, the objectivity of television news was highly regarded.

The most obvious feature of this huge audience is that it tunes in mainly between 5.30 and 10.30 p.m. each day, with the maximum number watching at about 9 p.m. This five-hour period includes five out of the eight weekday news bulletins. The main evening programmes of BBC1 and ITN therefore occupy commanding positions in the evening's schedule and each is regularly watched by about 17 per cent of the population, although this percentage falls during the summer months. For competing broadcasts such as the main and early evening bulletins, the audiences are about the same and there has been little change in recent years (see Table 1.1).

Seventy-five per cent of the viewers of an early evening bulletin will also watch one later in the evening. Since the main bulletin audience is larger, about 40 per cent of the audience of a 9 p.m.

Table 1.1 January–March weekday audiences amongst UK population (aged 15 and over)

		1965 %	1968 %	1971 %	1974 %
BBC1	9.15/9 p.m.	18	22	18	17
ITV	9/10 p.m.	20	18	18	17

Source: Annual Review of BBC Audience Research Findings, 1973–4

bulletin will have watched an earlier one. This degree of overlap is not acknowledged as a rule in the output. The easily noticed lack of reference back in the bulletins from day to day and week to week might indicate an assumption that the television audience is conversant with the general trends of the news; yet on a daily basis the daily repetition from bulletin to bulletin makes little concession to the overlapping of audiences.[3] Predictably, viewers of one channel are more likely to view that channel again than another channel, and the strength of the preference varies little for the three channels. Less predictably, Goodhart et al. found that the audience for news was no different in its viewing habits from the audiences of other programmes.[4]

The half to two-thirds of the population who watch one or more bulletins on the average weekday is described by both the BBC's and the IBA's research organisations in terms of socio-economic groupings using a simple market research classification. The composition of audiences as measured by the BBC is shown in Table 1.2. BBC is watched by a considerably larger proportion of the

Table 1.2 Estimated audience composition for television news (Wednesdays January–March 1974)

		Socio-economic group of persons aged 15 and over		
		A %	B %	C %
ITN	12.40	1	2	3
BBC1	12.55	1	1	1
BBC1	17.50	18	19	15
ITN	17.50	6	9	17
BBC2	19.30	1	1	-
BBC1	21.00	21	21	15
ITN	22.00	12	13	19

A = the 'top' 6 per cent in professional and managerial positions
B = the next 24 per cent, mainly white-collar and skilled manual workers, and
C = the remaining 70 per cent

Source: Annual Review of BBC Audience Research Findings, 1973–4

numerically small 'A' top socio-economic group than ITN. To a lesser extent the same applies to the 'B' group. The picture is reversed in the 'C' group, which has a larger following on ITN. This is confirmed in JICTAR research for the same period (March 1974), which shows that a majority of ITN's audiences was composed of viewers from lower socio-economic groups.[5]

Just as there is a larger proportion of ITN viewers among the lower socio-economic groups in the audience, so there is a difference in the way that the two organisations were viewed by those groups. In a 1970 BBC survey the 'C' group nominated ITN as their 'main source' of news more frequently than either BBC television, radio or the press. The 'B' group nominated BBC television more frequently than any other source. Only in the small 'A' group was the press cited more frequently as a 'main source' than television.

Although high levels of interest were claimed by all socio-economic groups, this 1970 study showed that a greater proportion of the middle class were 'very interested' or 'extremely interested' in the news. Where above-average interest was shown, however, radio and newspapers were more frequently cited as a 'main source' of news. A comparison of these 1970 results with those of the similar survey in 1962 shows that, in 1962, only 38 per cent cited BBC as their 'main source', 20 per cent ITN, 33 per cent newspapers and 17 per cent radio. In 1970 ITN had drawn level with the BBC. There has been a trend in favour of television news in general and ITN in particular.

The people interviewed in 1970 were also asked to say which source of news was 'most interesting'. Approximately equal proportions cited BBC and ITN (34 per cent and 35 per cent respectively), which again reflected a trend in favour of ITN since the 1962 study. But the most important result of these BBC surveys is to be found in what they reveal of the audiences' perceptions on the accuracy and trustworthiness of news sources. Results in 1957 and 1962 indicated that when asked to judge the trustworthiness of different sources of news, viewers elevated television news to a greater position of trust than either radio news or newspapers. These results were confirmed in 1970 (see Table 1.3). It is of course to be expected that the main selected source of news will usually be regarded as trustworthy. What is most interesting about this Table is that even among those who claim that the newspapers are their 'main source' of news, far more thought television news, especially on the BBC, to be more 'accurate and trustworthy' than the papers they read. The supposed lack of

Table 1.3 The 'most accurate and trustworthy' news

	Main source of news			
	BBC TV	ITN	Radio	Newspapers
Most accurate and trustworthy source is:				
BBC TV	75	18	19	39
ITN	5	59	6	13
Radio	5	6	61	12
Newspapers	3	5	8	27
Undecided	12	12	6	9

Source: Annual Review of BBC Audience Research Findings, 1973–4

editorial content on television news, the brevity of news items, the widespread feeling that 'the camera never lies', the lack of first-person statements, must all contribute to this result.

Trustworthiness is not, however, impartiality, and another slightly different question revealed that perceptions of the 'impartiality' of the television news have changed dramatically. BBC television news was thought to be 'always impartial' by 62 per cent in 1962 but only by 47 per cent in 1970. This trend was particularly marked among younger viewers. The BBC have suggested that this does not 'constitute proof of a decline in the BBC's standards of impartiality. It may be "society" that has changed, more people (particularly the young) taking a sceptical view of organisations seen as part of "the establishment" '.[6] This jaundiced view of the young cannot alter the fact that less than half of the sample interviewed believed the BBC was always impartial. To blame society for this is to argue without evidence that there is some platonic notion of truth obvious to and practised by the broadcasters, but only the audiences' increasing wilfulness prevents them from seeing it.

Concern at the absolute and relative decline in the standing of BBC news led to a further in-depth interview study in 1972 to determine the causes.[7]

At a general level opinions of the news cannot be separated from opinions about each channel as a whole. The news is not normally seen in isolation from the rest of the output; news watching is, according to these interviews and other studies,[8] habitual, often passive, and governed by behaviour patterns which have little to do with the news programmes themselves. These factors help to explain why public opinion emphasises stylistic differences; the BBC's formal, serious, humourless and establishment image, and ITN's interesting, friendly, human image. Among the attitudes to the BBC revealed in the 1972 study, the features singled out for criticism were a class bias in favour of the Conservative party and management, unnecessary complexity of language, and a stiffness and formality in presentation. On questions of detail (including the balance of items, the use of film, etc.) there is less clear differentiation between channels in the minds of the interviewees. Where it does exist it is more likely to arise from general attitudes towards BBC and ITN rather than from specific differences in news output.

Thus, for example, although BBC was thought to concentrate on more serious items and ITN to aim for a balance of serious and lighter items, this was a judgment about style rather than selections or balance of items. But despite these perceptions and the fact that the channels are supposedly in competition, our study found few structural differences between them. The amount of time devoted to each category of news; how long the individual items last; how much film is used, all vary very little from channel to channel. The main differences are that ITN industrial items tend to be longer than the BBC's, possibly a result of the different organisations in the two newsrooms. ITN have a team of three covering industrial and economic affairs, whereas during the period of our study the BBC had two correspondents independent of each other. BBC, with an education correspondent, cover rather more in that area than ITN. Aside from this, ITN uses more photographs than the BBC, and although they use about the same amount of film, ITN's film contains fewer interviews, fewer 'talking heads'. In fact ITN, certainly in *News at Ten*, covers a slightly greater range of stories rather more quickly than the BBC does. But none of this alters our fundamental finding that at a deep level, considering the range of journalistic approaches available, the bulletins are very similar. Out of the range of possible stories they both make a closely corresponding selection day by day, often down to running the same joke human interest items at the end.

In view of our own findings that, in most structural essentials, BBC and ITN news bulletins do not differ, the 1972 survey is significant. What the broadcasters know of audience perception has allowed them to turn these questions away from themselves and place the onus on the viewers. Thus, the BBC claims 'balance is important to the news viewer, but different sections of the audience understand different things by it. This being so, it seems unlikely that one could produce a news programme which seems "balanced" to everyone who sees it.'[9] By implication the broadcasters here again separate themselves from society, which is too differentiated to perceive the balance the broadcasters achieve. The achievement of balance cannot thus be checked against perceptions. What it can be checked against is not suggested.

In view of the findings that television news is still widely regarded as an 'accurate and trustworthy' source, it is not surprising to discover that, in this study, few people were concerned about bias or indeed believed that it existed. Those who did claim to detect deliberate or avoidable bias seemed to derive their view from an overall assessment of BBC and ITV rather than from a particular sense of bias in the news. Only some of the younger interviewees appealed to 'significant absences' as a source of bias, arguing that the lack of space given to the IRA or Vietcong point of view was evidence of a lack of objectivity.

Thus the BBC was able to dismiss the issue of bias as 'not really relevant', or at least simply a question of style. At the surface level, the results of our own study tend to confirm that there is little to choose between the two sources of news on grounds of content, technical competence and consistency. Yet by the broadcasters' own findings more than half the viewers now see their news output as not 'always impartial'. Whether those viewers are young, extremists, or otherwise, this change-around cannot be dismissed easily and has created what Anthony Smith, has termed the 'contemporary crisis in news credibility'.[10]

The news bulletins are becoming a contentious area. Yet it is in exactly this area that the appearance of credible neutrality is so crucial. For as Smith suggests, 'Credibility in the minds of the actual audience is the sine qua non of news. All else is propaganda or entertainment.'[11] Smith argues that the trades union movement has mounted a steady campaign against news reporting which fails to satisfy what trades unionists may feel are adequate standards of neutrality. Smith overstates his position, for with few exceptions the complaints about trades union coverage from unions are fairly informal and cannot be said to constitute a campaign. The notable exception to this is the pioneering study undertaken by the ACTT Television Commission in 1971. This study was an imaginative if albeit limited attempt to assess the impartiality or otherwise of news and other programmes' coverage of industrial and trades union issues. The researchers found that the BBC was erratic in its coverage, and tended to trivialise. It also suggested that on a number of occasions it had failed to maintain impartiality. This was said to be less true of ITN whose treatment, they found, 'evidences conscientious effort to maintain impartiality'.[12] It also criticised the trades unions for not being fully aware of the 'positive role they must play in supplying and checking television coverage of industrial affairs'.[13]

Smith remains convinced that the most likely outcome of a debate on any given area of news is that the 'news will in the course of time simply mop up the areas of discontent in order to regain credibility'.[14] However true this might be of the broadcasting institutions' on-going relationship with such bodies as the TUC or CBI, these political activities find little reflection on the screen. Smith appears to be wrong on both counts. The trades union movement continues to pay surprisingly little attention to television news coverage and our own research reveals that many trades unions still aim their information at the press rather than television. On the second count Smith was writing over three years ago, yet there are often instances of the cultural skewedness of industrial reporting such as those given in this and the subsequent volume. Moreover, unless the changes in news coverage are to be merely the outcome of pressure from powerful interest groups there has to be some process which provides safeguards for the less powerful.

It is an unfortunate lacuna in the work of social scientists who should be concerned with such important public issues that the job of monitoring and analysing media output has remained such an undeveloped area.

The literature to guide us in this area was virtually non-existent. The most relevant work was an American text, *Message Dimensions of Television News* by R. S. Frank who had videotaped and analysed seven weeks of American television news. This study, although very limited in scope and outcome, highlights some comparative findings which are worthy of attention; namely that network television news broadcasts reached a wider news audience than any other mass media source, and that television was described as the most important and credible source of news.[15] Frank explains this by virtue of the tendency of viewers to see television news as 'raw news' rather than as encoded or socially manufactured information. He was also able to demonstrate by comparing the airtime given to specific topics with Gallup Poll data that public concern with inflation and crime far outdistanced the amount of television news devoted to these topics.[16] The study concluded that whilst network news may not have necessarily changed or affected viewers' attitudes and values concerning specific issues, at the very least it reinforced perceptions of 'what is important'.[17] He therefore supports the long-held central thesis of much audience effects research, especially American, as to the prime media function.[18] But this ignores the crucial importance of the frames within which such reinforcement takes place. More recent work, both empirical and theoretical, has stressed the agenda-setting role of media as a clue to their importance as systems of social control.[19]

But it remains a commonly held notion that the media effect is reinforcing – rather than converting or casual. Academic psychology relies for its main causal explanations upon reinforcement theory, and demonstrates long chains of reinforcements can lead to changes in attitudes and behaviour. The further analysis of this paradox is essential if studies such as Frank's are to be of use.

The agenda-setting role, the ability to give certain events public prominence whilst ignoring others, is crucial in considering the news operation. Agenda-setting is not a value-free exercise, for as one academic has noted, 'Priorities in their [the broadcasters'] agenda tend to be set by the priorities assigned to topics or themes in the mass media. The informal daily education of the population is conducted by the mass media which tends to select some topics and ignore others, give precedence to some and not to others, and frame contexts and select content, all according to standards which perhaps owe more to custom than malevolent design, and more to unconscious synchronisation of decisions than to conspiracy.'[20]

The professionals tend to deny that their professionalisation has any real effect on the nature of coverage, arguing that within the limits of time and money their coverage is objective. But such objectivity assumes that the 'facts' exist outside of a frame of reference.

The work reported in this volume reveals that 'facts' are situated in dominant story themes, that such themes build upon basic frames of reference – basic assumptions about society viewed in particular ways – which often hinder the full and proper coverage of the events in question. Instead of admitting and attempting to overcome predilections for particular frames of reference, the television journalist tends to rely upon some notion of balance as a defence of objectivity. As one social scientist commented, 'by pairing truth claims or printing them as they occur on sequential days, the newsmen claim "objectivity" '.[21] It is this ritual bow to supposed balance which hides the fact that objectivity is not guaranteed by mere balancing acts.[22]

Our study indicates that professional and public faith in impartiality is indeed misplaced. The notion of cultural neutrality itself is only workable as an ideal, but in practice can never be achieved. The very process of cultural selection means that the news is not a mirror image of reality. The subtler and more routine assumptions of news production have rarely been subjected to the close scrutiny that this study attempts. The charge may be made that what is revealed via close study of the television news is merely that which is already known; that analysis of the output is really little better than the impressions which any careful viewer could carry away from the screen. Yet this study is not impressionistic. Indeed amongst the research team as many impressions or prejudices were refuted as were confirmed. Until the constant flow of television output is stopped, reviewed, and subject to close scrutiny and analysis of the kind here outlined then any charges of omission, slanting, or bias are merely hot air about a cool medium.

This book is intended as a step along the path to the systematic decoding of one centrally important element of contemporary culture. Contemporary cultural codes allow the often taken for granted generation of specific basic frames of reference. Such codes or routine handlings are not always easily revealed. For although they exist and are used in constructing and manufacturing news, they are so deeply ingrained as cultural assumptions that only occasionally, if at all, do they come up for questioning. It is only recently, for instance, that a public debate has emerged as to the proper length of the news bulletins. Whether news bulletins are readily comprehensible to most viewers still remains a virtual unknown. A Finnish study revealed that even with help from the interviewer 48 per cent of people questioned immediately after watching the news could recall nothing of the content.[23]

We would argue overall that many criticisms of the news offered up in the following pages are not easily resolved. There is much that is intrinsic in the social and normative coding of those messages we call 'news', which prevents the realisation of the aims of neutral communication of information.

The code works at all levels: in the notion of 'the story' itself, in the selection of stories, in the way material is gathered and prepared for transmission, in the dominant style of language used, in the permitted and limited range of visual presentation, in the overall duration of bulletins, in the duration of items within bulletins, in the real technological limitations placed on the presentation, in the finances of the news services, and above all, in the underpinning processes of professionalisation which turn men and women into television journalists.

Erving Goffman has observed that:

> Obviously passing events that are typical or representative don't make news just for that reason, only extraordinary ones do and even these are subject to the editorial violence routinely employed by gentle writers. Our understanding of the world precedes these stories, determining which ones reporters will select and how the ones that are selected will be told.[24]

It is clear that across a whole range of academic commentators there is agreement on the selected and non-neutral nature of professional frames of reference.

Stuart Hall has argued news values and the apparent neutral ideology of news production require that we examine the codes so that the ideology of news can be properly seen. Hall suggests that: 'News values appear as a set of neutral, routine practices: but we need also to see formal news values as an ideological structure – to examine these rules as the formalisation and operationalisation of *an ideology of news*.'[25]

The general claims of neutral news production are therefore undermined by much social science research. Indeed the pioneering and perhaps most important of early American social scientists in this field, Merton and Lazarsfeld, wrote in 1948 that the net effect of the mass media was dysfunctional for social change, 'for these media not only continue to affirm the status quo, but, in the same measure, they fail to raise essential questions about the structure of society'.[26] Yet such arguments are not limited to social scientists. Former professional media men and social historians have made similar points.

In the UK, television is an industry that operates upon government licence. Stuart Hood, a former editor of BBC News, argues that the financial control exercised by government is potent enough to force the BBC's policy into alignment with its general purposes. Government annually votes the proceeds of licence fees to the BBC and also retains the right to decide on the current size of that fee. Therefore, the overall ethos of the BBC within this political reality is an impartiality skewed as it were towards state policy – which makes sense of Asa Briggs's view, in his history of the BBC, that 'it is an organisation within the constitution'. Hood says 'in practice it is the expression of a middle-class consensus politics, which continues that tradition of impartiality on the side of the establishment'.[27] This ethos filters down as 'traditional wisdom' through the organisation, creating a situation where (in the case of news) 'a bulletin is the result of a number of choices by "gate-keepers". Each of the gate-keepers accepts or rejects material according to criteria which obviously, under no-system, can be based on individual whim but are determined by a number of factors which include his class background, his upbringing and education, his attitude towards the political and social structure of the country'.[12]

To the extent that broadcasting institutions are trading on some taken for granted consensus they are in the business of shaping public consciousness. The sources for this hidden consensus, as Stuart Hall has suggested, lie in the political culture of our society. The nature of the consensus is one bound up with parliamentary democracy. The prime values of British political culture become tied up with the institutions which legitimate its way of life. Perhaps it is for this reason that Lord Reith wanted the public broadcasting service to become a sort of value-free neutral administration, which is what indeed it has tried to become. But as Professor Tom Burns in his study of the BBC has indicated, the very 'rationale of the creation of a Broadcasting Corporation separate from the Government is that neutrality cannot be assumed in these regards. There is no culturally or normally neutral position to assume.'[29]

Historically this is particularly true with regard to industrial relations. After the 1926 strike the then John Reith, in a letter described by Asa Briggs as a document of basic importance, says: 'But, on the other hand, since the BBC was a national institution, and since the Government in this crisis were acting for the people, the BBC was for the Government in the crisis too.'[30] Reith goes on to say that: 'The only definite complaint may be that we had no speaker from the Labour side. We asked to be allowed to do so, but the decision eventually was that since the Strike had been declared illegal this could not be allowed.'[31]

Asa Briggs comments: 'There was no doubt, as an early message to the Station Directors put it, that there was a "certain natural bias towards the Government side". There was equally no doubt that the straight facts of working-class life were not well known to most members of the early BBC.'[32] Some researchers, such as Nicholas

Garnham, have suggested that this 'natural bias' is a continuing reality.[33]

These tendencies have not gone entirely unnoticed by other researchers, and a recent work on *Television and the Working Class* concluded that, 'the political function of television has generally been to promulgate and reinforce conservative social values in a number of forms including industrial relations, race, political protest and so on. Its broadcasting staff have been over-reliant on journalistic stereotypes and news values for its presentation of news and current affairs.'[34]

The notion that in representing some assumed consensus the BBC or commercial television is culturally neutral and does not serve to reinforce and shape some assumed consensus is a contradiction as Burns so perceptively notes in his study. This view of television and television news as political production or as an outcome of processes which themselves are political, has emerged in social science. Yet some cultural questions in media sociology are still very unclear. If much of the consensus is ideological; if the dominant ideas are the ideas of some ruling class or élite, if the dominant means of communication produce 'ideology', why does that ideology prove so slippery and intractable when under study? The phenomenon of mass communications provides us with a severe challenge.

Ideology is not something out there – independent of its producers. In part, ideology is the common-sense awareness of social processes. Common sense serves both to reveal and obscure what is going on. Most of what we or anybody else does is so rooted in our everyday practical activity that it is difficult, if not almost impossible, to bring our common-sense routine assumptions into view. Now common sense which, as the bedrock of social interchange, is the widest form of communication. It and public broadcasting systems, are all subject to cultural and class stratification; but broadcasting is coupled with a professional ideology and is tightly circumscribed by legal and conventional requirements.

Communicative power is about the right to define and demarcate situations. When we look at cultural power in this context we mean the power to typify, transmit, and define the 'normal', to set agendas. The power is used to reproduce highly selected events, and to manufacture news as if these events were the centrally important events of that day. In short, one must see the news as reflecting not the events in the world 'out there', but as the manifestation of the collective cultural codes of those employed to do this selective and judgmental work for society.

The year in which our study took place provides an interesting if rather crass indicator of the unquestioned cultural norms of our broadcasting institutions; for it was International Women's Year. Yet despite this, the national television news for all three channels was read either by one of 15 men or a person called Angela Rippon. This will come as no surprise to anybody, but it occurs in a nation where more than 50 per cent of the inhabitants are female. As with the news personnel so with news interviews: of the 843 named interviewees in the first 12 weeks of the year only 65, or 7·7 per cent of them, were women. The cultural conventions and codes of our society operate against a background of structures which regulate certain forms of cultural dominance. As the messages which television produces negotiate and pass through these structures, shaped by the relevant cultural codes, they cease to be neutral free-floating information about the social world.

We have concentrated on television news in our studies because it is here that supposedly neutral cultural production reaches its largest audiences on an everyday basis. Television news is concerned with the reproduction of information within the realms of a dominant consciousness.

The project has therefore had to concern itself with the vexed questions of cultural power and the consensual legitimation of beliefs. Culture, especially mass culture, is always in the process of change; if one wishes to be more than a spectator to such changes one must identify and map out the nature and output of one of the prime sources of communication. The kind of cultural decoding that reveals the systematic structure of day-to-day productions is needed. This work of decoding is an essential prerequisite if any acceptable theory of cultural production is to be forthcoming.

Thus far theoretical analysis of the mass communications industry has revealed that critiques which simply stress commercialism are in themselves too limited. Enzensberger, in his book *The Consciousness Industry*, talks of 'immaterial exploitation'.[35]

Enzensberger suggests that if you buy a book you pay costs and profit; if you buy a magazine or newspaper, you only pay a fraction of its costs; if you tune into radio or television your set is virtually free. Enzensberger's view is that the commercial exploitation of the media is not central and not intrinsic to it. According to him the consciousness industry's main business is to sell the 'existing order', to maintain the prevailing hierarchical pattern of society. Television news excludes that which falls outside of some assumed consensus and that to this extent it serves to render views outside of such an assumed consensus as irrational. Ralph Miliband has suggested that:

> The mass media cannot assure complete conservative attunement – nothing can. But they can and do contribute to the fostering of a climate of conformity – not by the total suppression of dissent, but by the presentation of news which falls outside the consensus as curious heresies, or even more effectively, by treating them as irrelevant eccentricities which serious people may dismiss as of no consequence.[36]

George Gerbner has said of mass communications that 'they are the cultural arm of the industrial order from which they spring'[37] – that is they reflect and reinforce the power structure of the society in which they operate. Thus they are *in all ways* political.

An important British study by Halloran et al. of the 1968 Grosvenor Square demonstration, followed the preparations for the event itself, in seven newspapers and in the programmes of BBC and ITV. They found that the event was anticipated as a stereotypically violent one and that the media borrowed one another's stereotypes; and the event was structured by the effect of the media on the demonstrators themselves, and most significantly on the police; and that even though the demonstration did not live up to its projected image as 'news', for it was relatively passive, the image of a violent demonstration persisted for many months after the event.[38]

The event achieves perceptual reality by being reported, while in addition consequences flow from the report which actually shape the original reality in accordance with the meanings given it by 'the news'.

Although, for instance in the buying of receivers and the paying of licences, it can be admitted that the mass media or the consciousness industry is in many areas highly profitable and is generally subject to the logic of commercialism, it does serve another and no less important function at the cultural level, a function which is unaltered by the private or public ownership of the medium.

This second function, the cultural legitimation of the consensus and the status quo is not subject to the narrow confines of commercialism. It is the role of television as a front-runner medium of cultural legitimation that is served by institutions of broadcasting however funded, whether privately or state owned. Historically broadcasting has always seen its role as one of entertaining, educating and informing within the confines of some unstated, taken for granted, consensus. But as we have said, this debate has not been a central issue in mass communications research.

The dominant media research tradition can be divided into two types of activity: audience research conducted by broadcasting organisations themselves chiefly for commercial or programming reasons, and academic research into production processes and output as well as audiences. The former tends to be extensive rather than intensive and is concerned with patterns of consumption rather than production. Unfortunately, ever since the BBC began systematic research in 1936, dominant interest has been to determine the size and the distribution of audiences. Today most of the resources of the Audience Research Department of the BBC and the commercial broadcasting equivalent JICTAR are taken up in this way although general audience reaction studies are regularly carried out as well as some specialised studies of individual programmes, such as we have utilised above.[39]

Our project was designed against a social science research background which has, typically, concentrated on broadcasting personnel and their audience and has not usually examined, nor indeed developed the methodologies to examine, television output itself. Since the screen reveals social communications of an extremely complex order, the creation of viable methods of analysis is a difficult but essential task; and one that can only be undertaken initially, we believe, by examining a very small area of output. What follows can be seen as illustrating the difficulties and potentialities of such an approach.

Our belief is that this approach ought to be at least as fruitful as previous work by social scientists which has concentrated upon determining the function and effects of broadcasting in our society. Unfortunately broadcasting organisations have not been active in this sort of area themselves; nor have they been encouraging to outsiders.

The project undertook a scientific critique of contemporary news output. It is a positive critique not in some arcane sense of limiting itself to checking or producing the facts but rather it is an attempt to document and map out the codes utilised in the practice of television news production. In seeking to assess, reveal, and demonstrate the results of this judgmental work we should stress that we are not engaging in negative criticism; but we are suggesting that by decoding, it is possible to show that the social and professional assumptions lead to particular frames of reference which are not neutral images of reality.

It is our assessment that most television journalists will readily admit that news stories, whether good or bad, are the result of much judgmental work. The positive contribution this study makes is by revealing some of the rules and structures governing the codings. The very notion of news values itself leads many researchers and commentators to question the inferential frames of news ideology. For behind these frames lies the task of interpreting and offering accounts of the central events of our days, months, and lifetimes. For instance:

> Good evening. For the first time in many months we have some good news. In the City the value of shares went up by about two thousand million pounds when the financial index jumped by 19·9 points to close this evening at 217. (ITN, 22.00, 24 January 1975)

What constitutes the definition of news or newsworthiness, may well be a more important question than the debate as to whether the output is 'biased' or 'objective'. For cultural bias is inevitable; however, its scope and direction are not. Hall has suggested that the common-sense constructs of the professional journalists guide the moral order and meanings which their stratified world presents and that news selection comes to rest upon an inferred knowledge about the audience and society which are limited by professional ideology.[40]

All of this is without prejudice to the defences of routine news practices normally presented. The durations of bulletins, the nature of the events included, the technological and economic limitations are not set in opposition to the inevitable cultural bias we believe to exist but are an expression of choices made as a result of those 'common-sense constructs'. News values and the expertise utilised in applying them are the mechanism by which events reach the screen in a skewed, mediated and highly selected fashion.

Galtung and Ruge writing on the structuring and selecting of news make the important point that 'the more similar the frequency of the event is to the frequency of the news medium, the more probable that it will be recorded as news by that medium.'[41] The orientation of television news is on a day-to-day basis. Events and actions which take time to develop and occur are likely to become news only at some culmination point. Halloran et al. in their previously mentioned study support the frequency observation by suggesting that 'a demonstration is a possible news event, while the development of a political movement does not have the correct frequency'.[42]

Frequency does not by itself explain the coverage of any particular area of activity on the bulletins. But with the other elements of the code it excludes or downgrades certain areas. Thus because of frequency, and basic assumptions about industrial life, the codes will contribute towards the newsworthiness of strikes rather than agreements, for example. But not all strikes will be covered equally. The process of selection operates to make some strikes more newsworthy than others. This newsworthiness is a function of the dominant view presented of the strike within the basic frame of reference which attaches to all strikes. The fact is that the sense of urgency engendered by professional practices leads to a spurious concentration on the immediate event. Today's news has no place for the 'late intelligences' that characterise the pre-telegraphic press. The result is that from day to day the bulletins seem to have been assembled by professionals who, while understanding broadly what events should be covered, are nevertheless rather amnesiac about the background and causes of any particular event.

Journalists insist that their professional integrity is based on a respect for 'hard facts', but definitions, within the newsrooms, of what constitute such facts are elusive. One major factor in determining 'facts' is the proven reliability of the source which increases the authenticity of the information in the mind of the journalist. Such sources would include the Government, recognised experts and press agencies. In everyday practice, as our observation work inside the television newsrooms shows, the exigencies of journalism and the inherited wisdom of the profession make journalists uncritical of certain sources. But research and verification and constant warnings on the laws of libel and precinct are part of the professionalisation of the young journalist. Thus formal legal and professional constraints operate to discipline the use of incoming information. This is especially true of television newsrooms where the dangers of editorialising, i.e. working without 'facts', are further enforced by the requirements of broadcasting Charter and Act.

The never easy definition of what constitutes 'fact' is further compounded in television journalism by the medium's ability to record actual events. Indeed a premium is put on such 'events' as an alternative to the head talking to camera which is, we shall argue, most of the news. An 'event', i.e. a good piece of film with action, stands a far higher chance of inclusion in a bulletin than a weight of unvisualised 'facts'. Yet beyond this, as it were, implicit faith in the adage that the camera cannot lie there is no body of professional wisdom comparable to the checking of sources and the constraints of the law against which to check the 'hardness' of 'events' coverage. Thus 'events' material is selected and used in accordance with criteria that have more to do with entertainment than the circumscribed caution exercised by journalists in their use of hard 'facts'.

In the industrial area the source of 'facts', our research suggests, tends to be management, this being in line with a general tendency to obtain facts from official sources. The labour side is looked to for 'events' – although as often as not this resolves into film of factory gate, picket line, demonstrations, *long shots* of meetings and little else. The balance, often achieved, of getting 'both sides' on the air is thus too crude. Time allocation to each side in a dispute is no guarantee of balance and impartiality. The concept of 'both sides' is often itself too limiting, and the coverage afforded to each side tends to be of a different order. Labour disputes are thought to involve 'photogenic discord'[43] and the source of the discord is more often than not the workers.

It is not the case that everything that is photogenic is made into news or that there is a dearth of alternative fact to give more balanced accounts. The routine practices of the newsrooms tend to look for 'events' angles in one area and 'fact' angles in other areas. In the reporting of industrial life these predilections are the result of the inferential frame used to account for the real world. Considering the complexity of television as a cultural form and in the light of the literature cited thus far we reject any crude notion of bias. The problems of producing a supposedly 'neutral' ideology render such notions unproductive. The central question thus becomes 'does television news as presently constituted help explain, and clarify events in the real world or does it mystify and obscure them?'

To move at least some way towards a systematic answer to this, we distinguished three areas of possible analysis. We sought other records of events in the real world apart from the television and we examined the linguistic and visual components of the television's account. Industrial coverage within the news, as indicated already by the examples given, is an exceptionally rich area for this work.

Industrial news is of extreme public importance. One central dynamic in British society, especially over the last decade, has been the conflict between what are often termed 'the two sides' of industry. In an hierarchical and profit-based society, it is unlikely that the central value system is neutral as between labour and capital. Although the value system of the broadcasting institutions is not, except at moments of extreme crisis such as 1926, exactly co-terminous with society's central value system yet it would square sufficiently in the industrial area for the problems of producing 'neutral' news to be highlighted.

Accounts of the real world

Much of the popular debate over television news has been concerned with the way in which it may distort, misrepresent and select or affect reality. Such questions are notoriously difficult to handle. Most claims of distorted presentation have been based on evidence which is circumstantial and subjective and therefore weak.

To try and avoid this we have taken the Department of Employment descriptions of industry in general and compared them to the television coverage. In particular the strike record of various industries in the DE statistics is compared with the number of television news items devoted to such strikes. This reveals a highly distorted picture of UK disputes during the period of the study. Television news emphasised strikes in some areas and ignored them in others, shifted its attention from one industrial sector to another irrespective of the continuing patterns of disputes in the first sector. Moreover the television news did not even cover all the disputes deemed to be of real significance by the Department of Employment.

Routine facts as to whether disputes are official or unofficial are rarely given, although most strikes are unofficial. Which trades unions are involved in a dispute is also not normally given. Another characteristic of industrial reporting is that despite the fact that the Department of Employment have a variety of causal explanations for industrial disputes these rarely appear on the screen. Basically, for television, the explanation is usually assumed to be money. Again this is strange, for a glance at the government statistics – well used by the television industrial and economic correspondents for other purposes – would reveal that between one quarter and one third of industrial disputes are about matters other than money.

Perhaps more serious is the spurious causality that often is implicitly created by the range and form of the industrial and economic coverage. For instance, despite much academic evidence on inflation which suggests wages only contribute a third to any price increase, the predominant assumption behind the coverage for our period was that of a wages-led inflation. We are not here arguing the merits of the case one way or another. We are simply pointing out that one interpretation is being used at all levels of news production to the virtual exclusion of others. The simple formula of demonstrating that price rises were being matched by pay increases during the period under consideration, conveniently ignores the fact that because of taxation, wage rises of 1½ per cent are needed to match retail price increases of 1 per cent and that in any case the relationship is not necessarily causal.[44] The one-sided causal explanation of inflation is a feature created by the news services' own range of sources for viewing this, yet is often presented as if it were otherwise e.g. 'Big wage deals are again blamed for rising prices . . .' (BBC1, 21.10, 29 April 1975); 'again' is the indicator of the narrowness of the range of views routinely presented. Much of this critique is of journalism itself and would apply at least in general to all media. So further work is presented which, by comparing the television coverage of industry with the press coverage, seeks to examine the reality of the particular constraints of television reporting. This is, as it were, a critique of the television journalists as journalists.

Linguistics

The almost complete lack of convergence between the discipline of linguistics, the literary and stylistic criticism of texts and the rag-bag of sociological content analysis provides an unfavourable climate for analysing news language. The pursuit of structure in texts by locating narrative units or episodes, and the distinction between form and content presupposed in most conventional content analysis have distracted from the importance of research on units of discourse larger than the utterance.

However, recent progress in conversational analysis has done much to show that the internal organisation of discourse, and the ways in which participants produce order in conversation, provide hope for a more systematic understanding. Ideally this is an understanding with a high degree of universality but at the same time a high degree of sensitivity towards particular contexts. The central questions in this kind of analysis are: how are successive utterances related; who controls the conversation; how are speaker turns organised; how are new topics introduced and old ones brought to a close; what type of utterance can appropriately follow what?

There is no doubt that Sacks, Schegloff et al. can claim some success in achieving a naturalistic and observational discipline that can deal with the details of social action rigorously and empirically.[45] There is also some evidence that the framework derived from conversational analysis may be appropriately applied to non-verbal interaction.[46] What is less clear is the relevance of conversational analysis to other uses of language and to public messages in particular.

The problematic of naturally occurring conversations may well give rise to descriptive apparatus of greater generality, but there are features of this kind of interaction between participants of roughly equal status which make it a particularly difficult form to begin with, for example very rapid and apparently arbitrary changes of topic; digression and the questioning of the questioner; ambiguity; misunderstanding, etc. Naturally occurring conversations are the most complex, subtle and difficult rule-governed form of spoken discourse. In contrast, the language of news (in its simplest, scripted form) is one of the most concise, and obviously stylised forms – and deliberately so. But this does not prevent the language of news from obscuring many of the processes of its production and reproduction. There is no doubt that methods of conversational analysis are appropriate to broadcast interviews despite their highly specific context. What is less certain is that they can be extended without large amendment to scripted news material.

The work of sociolinguists has been directed mainly towards correspondences between variations in language and variations in social structures (e.g. class, ethnic or sex differences). It is rare to find sociolinguistic work which goes beyond a consideration of the competence of speakers to communicate in varying linguistic and social contexts. Phonetic, grammatical or lexical variations are correlated with social context; or a language code is related to structural variations. Stylistics has had a similar concern with the possibility of a large number of stylistic 'registers' named after the field in which they are used (advertising, newspaper, scientific

English, etc.) and defined similarly by their grammatical and lexical features. Hence, for example, the 'register' of television commercials includes the following linguistic feature: the ratio of active clauses to passive clauses is 22:1.[47]

It is reasonable to suppose that news broadcasts have a register of their own which might have features such as: a high proportion of concrete nouns; a low adjective ratio; a greater than usual freedom in the use of pre-modifying nouns, e.g. 'The striking Newmarket stable lads' (ITN, 22.00, 20 April 1975). But this kind of work would suffer from a major limitation which is common in sociolinguistics as well as stylistics. A grammatical, lexical or phonological description of a text or series of utterances does not of itself show the function of the different elements in the discourse. Questions, for instance, need not only have an interrogative structure. They can also be imperative, declarative or moodless. According to Sinclair and Coulthard[48] this is only one of the aspects of language which indicates that the sentence, phrase or clause is not the highest order of unit which has to be studied.

For the moment let us return to Sacks et al., whose seminal work on the rules of conversational interaction provides us with a starting point for our own work on news talk. Sacks has called one such rule the *consistency rule* which is simply this: it holds that if some population is being categorised and if the category from some devices collection has been used to categorise the first member of the population, then that category, or other categories from the same collection may be used to categorise further members of the population. For instance we can find examples of news talk that run as follows:

> The week had its share of unrest. Trouble in Glasgow with striking dustmen and ambulance controllers, short time in the car industry, no *Sunday Mirror* or *Sunday People* today and a fair amount of general trouble in Fleet Street and a continuing rumbling over the matter of two builders' pickets jailed for conspiracy. (BBC2, 18.55, 19 January 1975.)

In this piece of news talk the category 'unrest' is used simultaneously to gloss such diverse phenomena as different strikes, short-time working, and a conspiracy case. The preferred hearing is clearly that we see (since we are talking of television) all of these as merely cases of 'unrest'. What this rule allows then is the following: that a hearer who uses the *consistency rule* and most of us do as a matter of course, will regularly not even notice that there might be an ambiguity in the use of some category among a group. In other words, where the *consistency rule* is in operation, by and large although that category or sentence is ambiguous, the *consistency rule* will be used by hearers so that they will tend not to hear any ambiguity with regard to that category or sentence. Various membership device categories such as 'strikers', 'militants', and 'shop stewards' are ambiguous. Each can have several distinct and quite different referents, e.g. 'a militant consultant', is not the same in any way as a 'militant miner', not because of 'consultant' and 'miner' but by virtue of the fact that the category 'militant' has changed. The regular operation in the media of the consistency rule means that ambiguities are glossed. In other words it says: hear it this way rather than another and do not notice the problems in using this category.

These hearers' maxims are the taken for granted rules which allow members of a given culture to make sense of everyday talk. We may pay attention to category devices which serve this purpose very well and there are some which have this essential property, and which are often used in news talk; namely, they are *duplicatively organised*, so that when such a device is used one counts not the number of 'ambulancemen', 'dustcart drivers' or 'electricians', but the number of 'Glasgow strikers'. That they are 'similar' by virtue of 'geography' or 'time' or that they often involve pay disputes then becomes the preferred causal explanation e.g. '*Still* in Glasgow, three hundred and fifty corporation dustcart drivers began a strike over a pay claim two months after a *similar* strike' (ITN, 17.50, 13 January 1975; our italics). The ideological glossing here, reflected in the deep coding of news talk, really implies that strikes are directly caused by unreasonable pay claims – a standard interpretive or inferential framework which tends to apportion blame for disputes on to the side of labour. So although it might be the case that the workers in Glasgow are on strike for a whole variety of reasons, if one uses a sentence which turns upon a *duplicatively organised category device* which has as its centre 'Glasgow strikers' then the hearers see any possible group of strikers as causal coincumbents. That is, duplicated categories are a case of devices in which the hearer's maxim is to hear it that way, rather than hear any ambiguity. Instances of news talk leading to spurious association and via such devices giving preferred and false causal messages are common and misleading, e.g. 'On a day when nearly 8,000 car workers were *made idle* by a *dispute* in the Midlands, the president of the motor manufacturers' society, Sir Raymond Brooks, has said no major British firm is making a profit' (ITN, 17.50, 1 May 1975; our italics). Competent hearers cannot really understand the above news sentence without taking for granted the assumed relationship or consistency between disputes and 'no profits'. One does not of course have to believe such an exclusive association – however the news sentence has to be heard as constructed around such a preferred connection. If we avail ourselves of present work on language we begin to observe and give instances of more general rules which are consistently followed and produced by television news. A central aim in the production of scripted news talk is to create preferential hearings which invite the competent listener to hear the talk as neutral. This often closes off any questions about evidence, about the problematics of such production and frequently rests upon unexamined causal inferences.

The main BBC1 news for 3 January 1975 provides a clear example of how supposed neutrality works.

> British Leyland said tonight they shared Mr Wilson's exasperation at the series of futile strikes within the corporation – *and there was more trouble today.* Two thousand workers at the Cowley plant near Oxford were laid off because of a strike by 350 men in the tuning department' (BBC1, 21.00, 3 January 1975; our italics).

The analysis of such talk requires little effort and almost no science. Especially if one remembers that Mr Wilson in his speech that day did not talk about 'futile strikes'. Rather he talked about 'manifestly avoidable stoppages', blaming management and finance as much as the work-force for any dispute. This in later bulletins is a framework to interpret motor strike stories which implies that they are irrational and caused by 'senseless' action on the part of the work-force.

Another rather different example can be given. We have found only one clear instance where a person demanding a wage rise was interrupted sympathetically. The person was Eric Moonman speaking for MPs. 'How much do you want?' asked Leonard Parkin. Moonman hastily replied that they hadn't had a rise for ages and that £1,000 was a minimum. He was going on to justify this when Parkin broke in with 'I would have thought that was very reasonable in fact' (ITN, 13.00, 7 March 1975). The lessons for researchers here are clear. If we wish to move beyond merely subjectively assessing such editorialising techniques as evidenced by the phrase 'more trouble' then we must learn to utilise the advances in sociolinguistics.

Visuals

Our study does not support a received view that television news is 'the news as it happens'. The vast majority of the coverage consists of men reading or speaking to camera in various ways; *talking heads* as they are called in the business. Only sport, political speeches, and other events deemed important enough to warrant the use of the Outside Broadcast unit which then broadcasts live into a bulletin can be said to be seen as they happen. But this is rare. And rare too is the extraordinary film that sticks in people's minds. The Turkish paratroops descending on ITN's Nicholson in July 1974 or the airline owner pulling men into the last plane out of Hué in April 1975 would be two examples. Most of the time, however, we get predictable film. We have noted that within this pattern working people come onto the news in a very limited range of circumstances which will be familiar. The factory gate is one essential location for industrial disputes coverage. As a result all those things which enhance a speaker's status and authority are denied to the mass of working people. This means that the quiet of studios, the plain backing, the full use of name and status are often absent. The people who transcribed our material have pointed out to us that the only time they had difficulty making out what was said was in interviews with working people. Not because of 'accent', but because they were often shot in group situations, outside, and thus any individual response was difficult to hear. The danger here is that news coverage is often offering up what amounts to stereotypical images of working people.

It is only film aestheticians who have exhibited any interest in this sort of problem and their work in this country in recent years has been heavily influenced by the French semiologists and the early Soviet art theoreticians.

'Semiology sets an idea of a text as an area of production of meaning, an idea of criticism which is to "follow the rhythm of the text". The emphasis is shifted from what the text means to how it means.'[49] In so far as this is a prime concern of the semiologist so it matches a major concern of the project. Talking of film another leading English practitioner remarks, 'Its work cannot be grasped by a simple inventory of codes, it poses analysis new tasks, a new objective.'[50] When faced with the problem of the analysis of the visual tracks in the news bulletins and their relationship to the linguistics of the audio track, and the significances of other sound tracks, we would agree with this. But in the actual work thus far presented as examples of how these objectives might be met we have found an unnecessarily complex use of language and a failure to integrate the description of the film text per se with the subsequent analysis of that text.[51] In addition to this problem, in this kind of work assumptions are made in an ad hoc fashion about the audiences, perceptions. But nevertheless, to utilise a number of the basic conceptions of semiology is obviously advantageous. It is after all the only agreed universe of discourse in the area, however much it might be in its infancy and whatever its problems. But we have also imported further concepts from the ethnomethodological area outlined above to help create an analysis whereby more systematic assumptions about the audience perceptions can be made. In essence this is to suggest that there are parallel 'viewers' maxims' to the hearers' maxims of Sacks and that these can be constructed by creating a 'lexicon' of news shots and determining the rules governing their juxtaposition. If such a methodology is to prove effective it must encompass the greatest variety of visual material used on the news, *viz.* newsfilm. In the second volume such a 'lexicon' will be attempted.

An examination of the system of news visuals can be used to draw up a list of viewers' maxims. Such an examination reveals visual expression of the underlying pattern of the assumptions of the broadcasters and their ideological base.

There are two essential logics governing the rules for juxtapositioning – for cutting – newsfilm shots. The first of these is a film logic. In the case of the newsrooms this is actually an extremely simple set of rules, coming from the basics of film grammar. In essence the permitted juxtapositioning of any shot is determined by the content of the shot preceding it. Thus, classically, a general shot is followed by a closer shot of something within that general shot or vice versa. If the shots are not related in this way there are two possibilities; first, the second shot is unrelated to the preceding shot because it marks the beginning of a new sequence or, second, the coverage of the material beginning with the second shot will within a few subsequent shots, which relate to it in terms of the main rule, link up with the material of the first shot. Put another way, in the classic narrative film the logic within the frame of each individual shot determines the content of the next shot. It is this logic that is used when film grammar is employed in a pure form by the newsfilm editors. There is nothing sophisticated about the way they use film – they seldom flash back; they never flash forward; they almost never move from shot to shot except by a straight cut. The rules governing how the interior logic should be applied are determined by the routine practices of the film industry around the world.

A trainee film editor will be taught a number of rubrics which govern the limitation of how shot B can relate to shot A when the material of both is the same but viewed from a different angle or at a different distance. First of the rubrics is that the angle of view cannot be too similar. To move from a long shot to a marginally closer shot can give the effect of jumping: it can be a *jump-cut* and destroy the logic of the sequence. The camera must not move from one side of the material to the other, otherwise the action will be reversed on the screen, i.e. a car moving from screen right to screen left will suddenly be seen to be going in the opposite direction if the camera is taken to the other side of the road for the second shot. Action from one shot to another must be matched. If the hand is reaching for the door in the long shot it cannot be seen to be already grasping the handle in the close-up and, similarly, if it has already reached the handle in the long shot it cannot then be still reaching for it in the subsequent close-up. Camera movement (i.e. pans etc.) must stop before a cut.

These are among the basics of narrative. They are altered subtly by time since they are dynamic. What is permitted – what the film editors assume the audience will not consciously see – changes. Tolerance for jumps has altered in the course of the past decade. Tolerance for the use of wide-angle lens has also grown. Tolerance for time slips, *flash-forwards* as opposed to the standard flashback have become so common that they are often these days used to engage the audience's interest in feature films which would, narrated 'normally', be too boring to watch.

Further, some of the old rules have fallen into disuse altogether. For instance an instruction to always cut a side view before a view from above is nowadays meaningless. There are many examples of this. But the essence remains. There is a corpus of rules, seen as such within the industry, which govern the way in which the basic rule as to the interior logic of one shot determining the following shot and shots can be applied.

But routinely in newsfilm this logic is not the dominant one. In most newsfilm the shots do not directly relate to one another in the ways we are used to from the feature cinema. Rather they are used to illustrate the audio-text and the rules governing their juxtapositioning come not from the visual but from the audio track – indeed largely from the commentary. Therefore they are filmed and cut according to an alien narrative logic – the journalistic. It is because the journalistic logic dominates the film logic that common professional opinion of television news journalists as film makers is a low one. 'Wallpaper' is how most of the output of the newsroom camera crews would be described. It is only in situations of danger or raw human emotion that the newsreel cameraperson achieves professional regard. These situations are not however the mainstay of the bulletins.

Of course the most obvious reason for this dominance is that the facilities necessary on location for achieving the logic of the narrative fiction cinema by definition do not exist for the news camerapersons. For instance the whole process of moving from long shot to close-up (or the reverse) using only one camera to record the scene obviously requires that any action in the scene be repeated.

Equally obviously, the repetition of action is something the news cameraperson cannot and would not go after, since it involves a level of intervention in the action not permitted by the observational role of such technicians.

In the area of current affairs and general documentary filming the main question at this moment of development in these forms is the extent to which the camera can intervene – and by that is meant exactly how close can a documentary crew come to treating their 'actual' material as a fiction director would treat his actors. For the newsreel cameraperson the routine story does not permit the luxury of this debate, never mind operating on it. The shop stewards' convener will not take the vote again so that the camera crew can get some close-ups. The terrorist will not obligingly appear at the window twice. The march will not restart. The workers will not re-enter the factory.

The structuration of newsfilm is therefore the result of an application of normal journalistic narrative practices as amended by the use of film logic where possible. As a result the viewers' maxims that can be generated indicate the extreme complexity of the messages and the skill required to uncode them on the part of the audience. For not only must the audience understand the simple rules governing the cut on the visual track and the culturally determined visual shorthand that makes up the contents of that track but also how that track is juxtaposed with the commentary.

For instance in classic film grammar things referred to in the commentary would be seen on the screen. In newsfilm they are often not seen. The journalistic structure might well be referring to absences – to things not happening – to abstractions. The film camera can only shoot objects that are there. As a result these objects are transformed by the process of juxtaposition into symbols. The empty station is not then simply a representation of an empty station but also a symbol of a rail strike. It is seen as such because of the juxtaposition of audio track.

This leads us to an interesting paradox. We would argue that in the bulletins such symbolic use of pictures, *prima facie*, might be least expected, because the bulletins report the world. Yet it is exactly the place in the non-fiction television output where such symbolic use most predominates. Indeed it is impossible to watch a bulletin without understanding the pictures continuously in terms of the symbolic structure created by the dominant logic of the journalistic narrative, even when film logic is simultaneously being used. There are indicators that in terms of comprehension the audience gain little from the visuals – they will understand almost everything from hearing the script alone.[52] However, that the visuals are not an essential carrier of meaning on the deeper level, has not been demonstrated to our satisfaction. The presentational devices in the news, such as the use of supercaptions, give to a constant viewer (and that is what viewers are in our culture) important clues for assessing critically the legitimacy of what is heard on the audio track.

Although the rules governing the juxtaposition of shots are therefore on the surface comparatively simple (as say when compared with a modern feature film) the relationship between sound and picture is extremely complex. The symbolic quality, itself a result of the logistics of news filming coupled with the shortness of item duration, are the cause of this.

Analysing news talk and visuals allows us to get at the preferred readings or seeings of a given item. It should be clear of course that the presuppositions of news production can only be revealed by close analysis of talk and visuals which will not only demonstrate the extent of the vocabulary employed but will show that although a large number of possible visual and verbal descriptions could be generated, only a limited number are consistently presented. The received notions of news selection – the gatekeeping function – are inadequate to the task of this sort of analysis. Selection at the linguistic and visual level are at least as important to an understanding of how the news works as a cultural form as are omissions and inclusions at other levels. Conscious of this, the project was designed to enable analysis on all three levels.

A window on the world – twenty-two weeks of television news

1 January 1975. The television news year began with the collapse of the Burmah Oil Company, a story dramatising and illustrating the mounting economic crisis which was to form the background to much of the reporting that we here analyse. Television news is a manufactured product based on a coherent set of professional and ideological beliefs and expressed in a rigid formula of presentation. The attentive and addictive viewer of the TV bulletins would have found nothing in the period of our sample from 1 January to 5 June (the day when the British people voted in a referendum to stay inside the European Common Market) which did other than confirm this belief system and conform to these presentational routines.

The substance of the research deals in particular with the reporting of the industrial life of Great Britain by both the BBC's newsrooms and ITN. But perhaps it would be as well to offer a brief impressionistic description of what any viewer might have carried away from the screen in general over these twenty-two weeks as a context for this industrial reporting.

The reporting of the economic crisis did not dominate the airwaves. As usual the surface confusion of the bulletins was maintained. They leaped, as they have done for the past twenty-one years of their existence, from economics to royalty, politics to crime. But such jumps are not random. They follow a rule-governed structure of presentation.

It is one of our major purposes to reveal and describe such rules which govern the routine presentation of television news; but there are other routine practices, equally rigid and less easily analysed, which also affect the deeper structure of the news. In the reporting of the economic crisis the television newsrooms adopted a number of strategies for concretising and dealing with economic affairs, most of these also adopted by the press. Central to these strategies was a constant and ongoing assessment of the government's agreements with the trades union movement – the Social Contract. This document set out guidelines for collective bargaining in the period after the previous administration's attempts to impose legal wage restraint. The containment of wages was presented as a key to the control of inflation and many of the agreements signed by workers were reported by the newsrooms as being either 'within' or 'outside' the guidelines of the contract. Indeed, often a major criterion operating for making the signing of such agreements news was the very 'breaching of the social contract'.[53] Thus millions of workers who settled their 1975 wages levels without dispute were reported in these five and a half months – the biggest coverage being given to the miners; but postal workers, construction workers and some sections of the printing industry, the railmen (with the exception of some signalmen), and the power workers, were also reported.

The major disputes occurred, according to the television news, in the car industry, at the London docks, in Fleet Street, in connection with some rail signalmen, mainly in England, and a 'wave' of municipal strikes in Scotland. The major car industry strike, highlighted by a speech made by the Prime Minister immediately before it occurred in the first week of the year, was by engine tuners at British Leyland Motor Corporation's Cowley plant. This was seen as an instance of the 'manifestly avoidable stoppages' mentioned by Mr Wilson. In common with the dockers, who were striking against containerisation in the ports, and the National Graphical Association printers, who wished to maintain a forty-pence differential against other print workers, these disputes were to do with erosion of skilled status, regrading or the introduction of new techniques.

The strikes in Scotland eventually narrowed to the 'health hazard' created by the municipal dustcart drivers in Glasgow which resulted early in March in the troops being called in to clear the refuse.

The other major area of dispute was the Health Service where throughout the period doctors at all levels in hospitals adopted various trade union strategies to obtain new agreements and working conditions. At the top end, the hospital consultants' argument with the Secretary of State for Social Services became coloured by a congruent debate as to the rights for private medicine to be practised in parallel with the National Health Service. This debate called forth industrial action against private patients by auxiliary hospital workers in some areas – notably Wales and the Wirral.

All these conflicts and strikes took place against a background of rising unemployment (routinely reported) and recurrent stories of layoffs and short-time working, with particular and perhaps undue emphasis being placed on the vehicle building industry. Here industrial reports were sometimes coupled with economic reports of world-wide falling demand for vehicles – yet often in such a manner as to suggest some causal connection between strikes, output and sales. Steel and the British Steel Corporation's plans to shed jobs caused Minister of Labour Michael Foot considerable personal embarrassment because the plans deeply affected his constituency Ebbw Vale. Cutbacks announced in May for the Scottish steel industry provoked less coverage.

Against this background the period was also characterised by stories of business bankruptcies and failures. Apart from Burmah Oil, the general reporting of vehicle building troubles paved the way for the government's subvention in British Leyland and Chrysler, outside our period, and the continued attempts to save Aston Martin at the beginning of it. Other attempts at government subvention in industry were dealt with as political stories; for instance, the final decision to support the workers' co-operative at the Meriden Motor Cycle Works. These subventions often involved the Secretary for Industry, Tony Benn, by this time the most controversial figure in the government. The closure of Imperial Typewriters and the investigation into the affairs of British Leyland, the government acquisition of 62 per cent of Ferranti's, were typical of the way in which the political aspects of industrial stories were played. The Stock Market and the fluctuations of the Pound were routinely reported, a mini stock boom in January being the only occasion when this area of activity was treated as a discrete story and placed high in the bulletins. Thus the problems of the Pound – it fell from \$2·34 to \$2·31 in the sample period – were not over-played.

The relationship of government to industry was further stressed by meetings between the Confederation of British Industry, the Trade Union Congress and the government being fully reported. In Parliament itself, the introduction of the Bill to establish a National Enterprise Board – the plans for nationalising the aircraft manufacturing industry, engendered major political rows. The Minister for Industry was so much identified with this strand of stories that in some cases (e.g. the collapse of Imperial Typewriters) he was associated with workers' sit-ins and demands for government aid as soon as the story broke.

Of the major political events of the period the resignation of Edward Heath as leader of the Conservative Party and the election of Margaret Thatcher as his successor was one of the few stories to be treated at such length as to swamp the bulletins;[54] that is, it was given such prominence on the days in question as to preclude coverage of much else. Apart from the novelty of a woman coming for the first time to the leadership of a major political party, the process of appointment was also new, an open election among the Conservative MPs instead of the 'emerging leader' system of appointments of the past.

The other story with the excitement of a general election campaign was the referendum debate which reached its climax on 5 June and marked the end of our recording period. The question of the UK's continued membership of the Common Market cut across party lines and 'pro' and 'anti' Common Market organisations emerged to fight the battle. Apart from the full reporting of the activities of these two groups, attendant party battles were also well covered, notably the Prime Minister's decision to allow members of the Cabinet to break ranks and speak as their consciences dictated, and the different sides taken on the issue by Mr Wilson and the National Executive Committee of his party. Other rows centred on the arrangements made for counting the vote, the 'anti' group insisting that it be counted by constituency and winning the point.

The referendum campaign did not affect coverage of Common Market affairs to the extent that might have been expected. The EEC Summit in Dublin at which the Prime Minister and the Foreign Secretary finally hammered out the revised terms of our membership, was seen by the newsrooms as much an Eire story as a referendum debate story. The agreement over sugar secured by the Minister of Agriculture and the row about the EEC Regional Development Fund were the other running Common Market stories during the period.

Less major political stories included the debate about the Trade Union and Labour Relations (Amendment) Act, again involving Mr Foot, which came to centre on the rights of the NUJ to organise newspaper editors. Like many stories affecting the press itself, the row over the publication of the late Richard Crossman's diaries being another, this was fully covered although it did not reach its climax until the autumn. The Queen's income, the Civil List, was also a controversial subject. The Royal Family were particularly active during the period with state visits by Her Majesty to the West Indies, Mexico and Japan and with Prince Charles in the Canadian Arctic and Nepal for the wedding of its king.

Britain's relations with the world, apart from the European Community, did not dominate the news. During this period the Prime Minister went to both the USA and the USSR. The Foreign Secretary visited Africa and contributed to a small but perceptible movement on the part of the regime in Rhodesia. Mr Smith, prodded rather more effectively by South Africa, held talks with some Zimbabwe national leaders. A row occurred over the visit to the UK of Soviet Trade Union leader Shelypin, as a guest of the TUC, but other visits by the Prime Ministers of Australia, New Zealand and Canada were treated as being of minor importance. The Commonwealth Conference held in Kingston was also not over-emphasised, but was covered. More play was given to the misdeeds of the British ex-chief of the Hong Kong police Godber, and the lengthy process of the removing the run-away MP John Stonehouse from his Australian haven. Indeed the Stonehouse story received such emphasis that he appeared on the news screen in the early part of the year more often than any other figure.

In Cyprus, in the aftermath of the Turkish invasion, there was a general hardening of the lines with the Turkish Cypriots declaring an independent state. Britain was involved in airlifting to mainland Turkey some families caught in the British sovereign bases on the island, which provoked riots against the British Embassies in Nicosia and Athens.

In Europe the Danish election was reported and occasionally there were stories of economic difficulties such as the one-day Italian general strike. Portugal was the major European story. The riot at the social democratic congress was fully reported as were the attempted counter coup by General Spinola in March and the elections won by the socialists under Soares in April. Apart from this, the newsrooms fully documented other aspects of the Portuguese revolutions, filing stories on everything from agrarian reform to the threatened presence of Soviet ships in Portugal's NATO ports. The situation in Portugal's African ex-colonies was less eventful during the first half of the year.

The OPEC summit held at Algiers was treated as part of a strand of stories about oil which included the UK government's plans for a stake in the North Sea oil operation (eventually favourably received by the oil companies), and the problem of recycling petro-dollars (sometimes concretised as tales of sheikhs acquiring huge chunks of London).

In the middle east, Kissinger's shuttle diplomacy, leading to a rejection of his terms by the Israelis in March, was the major event. In Lebanon the signs of the communal strife, which towards the autumn was to develop into a civil war, were seen with first reports of tension in mid-April. King Faisal was assassinated by his nephew but there was a smooth transfer of power to his brother and the assassin was executed without upheaval. In Formosa ten days later Chiang Kai-Shek died and an equally smooth transfer of power brought his son to the leadership of the Nationalist Chinese.

A major war continued in Ethiopia around Asmara. Although this was not at all fully reported, the Eritrean Independence movement appeared to have suffered a major defeat at the hands of the Ethiopians. The Kurdish rebellion in Iraq petered out, again without full reporting, by early April.

The major foreign story was of course the final victory of the communists in Cambodia and South Vietnam. The routine stories of communist attacks slowly gave way after the first ten weeks of the year to greater coverage as the advance gathered momentum. The problems of the refugees and the sudden removal by air of large numbers of 'war orphans' coloured the coverage. The fall of Saigon on 30 April was, like the election of Margaret Thatcher, one of the swamping stories of the period. American reaction was covered but typified by President Ford's running for his plane to avoid answering reporters' questions on the subject.

American domestic affairs were not emphasised, the economic problems of the USA, and Ford's plans for dealing with them, being scarcely reported at all on the screen. Greater coverage was given to the continuing strand of Watergate and Watergate-related stories, especially the revelations in Congress about the activities of the Central Intelligence Agency. President Ford's visit to China received routine coverage and the *Mayaguez* incident, in which the USA compelled the new government in Cambodia to surrender an American ship, was also given a major play.

In common with other years in this decade, events in Northern Ireland formed a substantial strand of reporting. During this period, the toll of bombings and deaths in Northern Ireland continued and was routinely reported. The IRA truce was the major political event of the period but it seemed to have little effect on the killing. There were bombings in England too but some, for instance those at London hotels, were not attributable to the IRA. There was a bombing in Scotland claimed by the Tartan Army. The main Ulster stories, apart from the truce, were the trials of those accused of the bombings which occurred in 1974 at Guildford and Birmingham.

There was only one major demonstration during this period, the three-week march to London in support of the jailed building pickets – the 'Shrewsbury Two'. There were other demonstrations – textile workers, egg producers and herring fishermen all staged protests to draw attention to the effects of what they claimed were unrestricted imports of their products. The fishermen's blockade of their ports followed similar action by their French counterparts earlier in the year. In the last week of February the Scarman Tribunal found the International Marxist Group responsible for the events in the previous year in Red Lion Square when a student was killed.

The first attempted hijacking on UK soil took place in January and swamped the bulletins, unlike similar events in other countries, such as the bazooka attack on an El Al plane at Orly Airport in France during the same month, which received routine coverage.

The final story to swamp the bulletins was the first major Tube disaster ever in London. The death toll reached 41. A coach crash in Yorkshire killed 32 in May and heralded a spate of similar stories. Consumer affairs were largely a matter of ever-rising prices; large increases in the television licence fee, postage rates, electricity and gas charges and the continuing struggle over the price of food, were all reported.

These were the salient events of the period under study as seen by the television newsrooms. In addition, as ever, a succession of pools winners, honoured citizens (Sir Charles Chaplin, Sir P. G. Wodehouse), obituaries (Sir Julian Huxley, Sir P. G. Wodehouse), sports reporting, and jokes concluded the bulletins. Fiji miners demand extended break for lunchtime sex; the managers of Tiffany's New York forget to lock up but nevertheless get nothing stolen; house falls into Mersey – nobody hurt; and a little crime – a strange case of exorcism and murder in April, but notably the protracted search for heiress Lesley Whittle resulting, after TV offers of ransom money, with the discovery of her body on 7 March, made up the rest.

We shall argue that this highly selective mixture nevertheless presents an entirely coherent view of the world at a deep level. As

Stuart Hood has pointed out, at the end of the major bulletins, the newscasters almost always shuffle their scripts into a neat pile and cap their pens. Order is imposed. The ungodly continue ungodly but the audience can sleep easy – the window on the world reveals an unchanged and unchanging view.

2 CONSTRUCTING THE PROJECT

> If the editor's yardstick is not a narrowly mathematical one, then neither can a purely numerical index be the measure of the honesty of his intent or of the soundness of his judgment.
>
> Gerard Slessinger, Managing Editor, BBC Television News

The basic design

The television image is a manufactured one, involving processes of extreme technological refinement and relying for its communication effectiveness on complex systems of culturally determined message coding. The manufacturing process applies as much to the output of non-fiction television news and current affairs as it does to light entertainment and drama.

But the ideology of news producers and reporters is one that lays great stress on neutrality. This is expressed in a series of professional constraints and practices backed by a commitment to impartiality and balance imposed by law and convention. Obviously events are not transmitted raw and untouched by human hand to the audience, yet the professional ideal of news reporting implicitly assumes something like this: that whatever processing is involved in production it does not alter the essential 'truth' of what is reported. How then does this conflict between manufacturing process and the ideology of neutrality express itself on the screen?

Since there is such a range of news and current affairs output, in seeking answers to the above, two strategies were possible. We could look at the whole output for a short period or one particular area as defined by content over a longer period. It was felt that an examination of one area over a longer period would probably reveal more of the routine practices and assumptions governing the production of news and current affairs and thus better enable us to see any regular features of the manufacturing process and the producers' ideology. We have indicated above which particular area of news was considered likeliest to yield results.

In recent years there has been growing criticism from both sides of industry as to the nature of television news, specifically questioning the narrowness of its industrial interests and the way it routinely simplifies complex industrial issues. The area of conflict between management and labour, as we have said, highlights the problems of maintaining an ideology of neutrality while reporting this area. The 'One Week' study revealed the possibility of analysing this industrial reporting and showed, albeit crudely, how tenuous this neutrality was when the output was subject to close monitoring.[1] In addition to this background of public debate there is statistical material emanating from government about industrial conditions as well as whole areas of university research on management, workers and industry. Above all, industrial life is a continuing major source of news and was chosen as a topic for study for all these reasons.

Projects of this type have always faced one major problem: the output of broadcasting is constant and, by nature, ephemeral. It only recently became possible to store broadcast material in such a way as to make it easy of access for scientific purposes. The video-cassette machine is both extremely easy to operate and costs, when compared with the video-tape machines used in broadcasting, very little. This device therefore made the project technologically and economically feasible. It was also hoped that the automatic timing facility that certain types of video-cassette recorder (VCR) have, would help with the considerable work problem caused by the need to monitor long hours of the television output.

In 1975 there were eight news bulletins every weekday, some half dozen regular current affairs weeklies (although not all were produced by current affairs departments), and three current affairs magazines on a more or less daily basis. At the weekend there were further regular bulletins and one more current affairs programme.

These broadcasts occurred throughout the day – from lunchtime to late evening. We decided to record onto video-cassettes one year's output of industrial news which was to include within all those programmes recorded all items concerned with the TUC, all strikes, other forms of industrial action, general trade union activity, reporting of the CBI, business and economic affairs, business successes, industrial accidents, labour and economic legislation. A year was chosen to take account of seasonal variations in both industrial life and television programme patterns.

At the design stage of the project three phase semerged: first, a period of three months to set up the machinery (six VCRs), establish recording techniques, recruit the research team and design basic content analysis sheets which in turn involved the generation of a basic categorisation system; second, a period of one year beginning 1 January 1975 to record the programmes; third, nine months to finalise analysis (for it was envisaged that analysis work would also be done during the second period) and to write this study.

Since there is virtually no literature in this area it was extremely difficult to know whether such a design would be viable. We had no clear idea of what amount of pertinent material would emerge on the screen during the year. We estimated that 240 hours of video-cassette should be sufficient to store this area of output over one year. As described below we were wrong in our estimates of the material which would be pertinent to the study and how much coverage it would receive.

The initial phase – September to December 1974

The new technology

There were two major problems: bedding down the equipment and establishing our own methods for using it involved collective training, since only two of the team were familiar with VCRs. Second, and more important, there was the designing of the basic story category system which also involved determining what range of observable data could be encoded across the whole of the output so that subsequent comparative work could be done with the industrial material.

Because of the built-in automatic timer and their relative cheapness, Philips VCR N1500s were chosen. The VCRs gave us considerable difficulty. Of course they are essentially a domestic appliance and not designed for the level of usage to which we would have to subject them, but even at the outset they were unreliable – one machine having to be cannibalised to keep the others going. This meant we did not have all of them in service until Easter and they were still subject to breakdown. We decided that the work would require six VCRs – three tuned permanently to the three television channels, two reserved for editing and viewing purposes, and one spare. In the event during phase two we were sometimes reduced at the end of an evening, through malfunctions, to only one fully operational machine. A considerable portion of the project technicians' time was taken up in maintaining these machines and this was especially true during the first phase.

The university's television service provided excellent backup facilities for the project. For instance, each of the three VCRs tuned to the broadcast output was placed in a custom-made stand with a televisions receiver mounted above it. The resultant bank of three VCRs tuned to BBC1, BBC2 and ITV respectively, with three televisions above them, made the actual work of noting the contents of the news very much easier than it would otherwise have been.

Further, a system linking all the VCRs to a master VCR so that video-cassettes recorded on them could be transferred onto an archive video-cassette (AVC) was built. This was designed to enable us to edit the industrial and related items out of the coverage and thus build an industrial archive on video-cassette.

It was this editing requirement – an inevitable result of the long-term strategy adopted by the project – that lay behind the problem of basic story categorisation. The essential factor here was to work out what observable data across the whole of the output were worth noting for the purposes of subsequently comparing such data between different areas of news stories.

Obviously we would have to edit the output down quickly so that the store of video-cassettes would not be expended. We calculated that some 1,600 hours would be recorded during the year. We anticipated keeping only 15 per cent. Editing decisions would have to be made while it was still difficult to determine what would be of use in the third phase of the project. As a preliminary it was therefore essential to generate a category system for all news stories.

Story categories

Rejecting the generation of categories according to *a priori* analytic criteria and following the advances of ethnomethodology we decided to take 'common-sense' criteria which stemmed from what we then knew of the professional practices of the producers. This, we felt, was preferable to the imposition of our own analytic categories which, however finely developed, we decided may not necessarily relate to actual newsrooms. We were, after all, studying the manufactured artifact and its processes of production and the divisions related to those practices and not some abstract analytic.[2]

In the newsrooms the basic discrete areas of reporting are represented on the screen by different correspondents – diplomatic, industrial, economic, etc. This division of labour is not hard and fast. We noticed, for instance, that the economic correspondent would often be reporting industrial news of interest to the project. Some areas are discernible as story-types without having correspondents allocated to them – human interest and disasters are areas of general reporting, for instance, which routinely occur. Thus the system outlined below was not totally delimited by the presence of an identifiable correspondent but by strands of stories that a professional would readily recognise as such.

The initial list of categories therefore broke down as follows:

10–11	Parliamentary politics	40–41	Economics, City
12	Northern Ireland	42	Economics, currency
13	Terrorist	43	Economics, commodities
14	Demonstrations	44	Economics, business
15	Labour Party internal	50	Crime
16	Conservative Party internal	60–61	Home Affairs, local government
17	Liberals	62	Home Affairs, consumer
18	Other parties and pressure groups	63	Home Affairs, quality of life
19	Politicians personal	64	Home Affairs, race
20	Industrial	70	Sport
30–31	Britain and the World	80	Human interest
32	The World	90	Disasters

A detailed description of the basis of this categorisation system and how it was operated is given in Appendix 1.

There were three basic errors in this list. This can be expressed as three correspondents being forgotten. First, education was put with local government. Thus tertiary education was not included as a separate category and was therefore classified as 61 with other education stories so that 61 became in effect, local government and education. Second, defence was entirely omitted at the initial stage. This was realised early in the second phase and eventually during the process of cross-checking the categorisation, a defence category

(65) was added and all defence stories which had been placed *faute de mieux*, in science (00), human interest (80), or environment (63), were taken out and placed in 65. Most religious stories were classified as human interest and were left in the category 80.

The basic assumption that the project adopted as regards the problem of categorising the material was that in the minds of the men and women who made the news, all categories were in fact dynamic and reflected the range of stories currently being reported. Thus a decade ago such a list would not have included Northern Ireland as a discrete area of reporting. As it was, we found that race (64), and demonstrations (14), were not so prominent as we had assumed on the basis of our collective memory of the news. Further, in terms of actual reporting the categories are by no means fixed, as the angle taken on a story can determine which category it ought to belong to. So stories have a daily dynamic which moves them from category to category as they develop.

Sugar is perhaps the best example to be taken from the data recorded in phase one. At times this was presented as a consumer story (62); as an industrial story (when the sugar refiners went on strike) (20); as a Britain and the world story (when the Minister for Agriculture bargained over sugar prices in Brussels) (31). The basic principle we adopted was that each project member logging a particular news bulletin should categorise according to the main thrust of the story, changing if need be, from bulletin to bulletin, and that the problem of maintaining consistency would be dealt with by a system of group checks, and the use of the computer to check for individual variations. At base the project began with the assumption to which it still adheres that too strict a regulation of the process of categorisation would ultimately involve serious distortions of the data to fit the categorisation system itself. We felt that any errors due to subjectivity were preferable to this and that the very range of the sample would minimise such subjective errors. A final revision revealed that only 4·2 per cent of all items in all bulletins were penumbral and required recategorisation; and after this reallocation no one story category was altered by more than 16 references.

Another major problem was itemising the contents of the news. British television practice is to try and place as many related items as possible into coherent packages within the bulletin.[3] So here again we decided to follow what we knew of journalistic practices rather than trying to establish an *a priori* analytic categorisation system which would not necessarily reflect those practices. The project therefore developed a tendency to place separate juxtaposed reports on the same topic into one item, especially if, in the two-newscaster situation of the main bulletins, the whole series of reports was linked by one man. A change of newscaster often indicated that the second story segued out of the first. It was marked therefore as a separate item. For example news from Northern Ireland and the Republic would normally be one item with the possibility of segueing into an English bomb trial.

As with the story categorisation problem, the project assumed that the dangers of errors were to be minimised by a system of group checks. Then, moreover, computer cross-checking was established with the added factor that we acquired in the course of our observational research a certain number of running orders from the newsrooms against which we could check our own itemisation pattern.

Observable data

A similar reliance on what was known of the practices of the newsrooms was used to generate an initial logging sheet. In determining what observable data would be recorded for the whole of the sample, we assumed that certain decisions by the news producers revealed their perception of the relative importance of one story as against another. Obviously these included the duration of any given item and its placement in the running order of the bulletin, the longer the item and the higher in the bulletin, the more important it would be said to appear to the newsrooms and their audiences. Further, we assumed that if there were different patterns of technical input (e.g. film) from story category to story category this would reveal the different strategies adopted by newsrooms with different story types. But within each category we assumed that the more inputs of a technical nature into a story the more important it could be said to be since each input represents a decision to commit effort, time and money in one direction rather than another.

Apart from being a record of the data as would yield such clues, the basic logging sheet would also have to act as a record of all those stories which were not retained on the archive video-cassettes (AVCs). Therefore it was also necessary to note such things as who appeared on the programme, both television personnel and interviewees. Between September and December the initial logging sheet went through four variations before it was printed by the university. In its final form it divided into three sections: basic duration and placement information with story description and categorisation; people appearing; and technical inputs (Figure 1 in Appendix 2). A detailed description of this sheet is given in Appendix 2.

The logging sheet was designed for all output, news and current affairs, but the contents of the news bulletins were seen as being the most complex simply because the duration of items within a bulletin and the range of technical inputs was greater than with either the weekly programme or the daily magazine. These would use the same range of inputs spread over greater durations. Thus it was felt that if the logging sheet could accommodate the range of data involved in the bulletin it would also accommodate programmes in the current affairs area.

The logging sheet emerged after some four drafts, the essential difference between them being that they gradually simplified what could be noted – establishing, for instance, the codes for interviewees and graphics. In the first drafts of the logging sheets we thought to include timings of film inputs within the items but after the second dry run at using the logging sheet the work load involved prohibited that. It was decided that at some point in the second phase a complete set of bulletins would be kept and not edited down. Basically this was to enable us to compare the language of other areas of news reporting with the reporting of industrial life and for comparative analysis of the visuals. Since this was the case, it was decided to leave the determination of average lengths of newsfilm until that stage of the project and not to attempt this with the basic logging sheet.

Similarly the noting of the presentational device of using pictures behind the studio personnel's heads, a process using colour separation overlay (*Chromokey*), was also abandoned after the first drafts of the initial logging sheet. It was decided to note this also when the visual analysis was undertaken using the archive of complete news bulletins. One presentation device – electronic screen-writing techniques – was ignored throughout the course of the project.

In so far as the project was able to begin recording on 1 January 1975 with the sheet described in Appendix 2 which, in practice, required no subsequent revision, and in so far as the editing system did establish an archive, again without revision, this aspect of the initial phase of the project was a success.

Computer coding design

In the course of developing the initial logging sheet and before data collection began on 1 January it had become clear that the numbers involved in a description of the general characteristics of each bulletin would be very large, and an experimental analysis of a week's news demonstrated that such a task, if done manually, would be almost impossible to sustain for a long period. Although no computer programme had previously been developed for this kind of analysis, this problem was one clearly suited to processing by computer. As a result of using the initial logging sheet, this analysis would involve managing a mass of news items differentiated in a number of rather complex but well defined ways. Rather than write a new programme for the specific purposes of the project, it was decided to make use of existing facilities, namely the Statistical Package for the Social Sciences[4] widely used in social science research. This is an integrated system of computer programmes designed to include a number of the most commonly used procedures in social science data analysis. Thus in addition to descriptive statistics and frequency distributions the package contains procedures for cross tabulations, correlations and factor analysis. Many of these subprogrammes were used in subsequent analysis.

The transformation of the initial logging sheet into machine-readable form was made in two stages for the sake of accuracy. Primary log data was first reduced to numbers and was entered onto a standard coding form. These records were then punched onto cards. Although the coding was designed to ensure that the minimum of information was lost in this procedure, there were two main areas of 'leakage'. Since there was no way of classifying stories beyond the simple story categorisation, and since headlines would take up too much space on a punched card, the information contained in the headlines and subheads was lost for the purposes of computing. Loss of information also occurred in the measurement of inputs. Whereas the initial log contained information on how many pieces of film (for example) were used in a single item, the computer coding reduced this to a simple question of presence or absence of film.

Apart from these omissions, it was possible to include all the remaining data for a single item on a single punched card. The small numbers on the initial logging sheet (Figure 1, Appendix 2) reference the corresponding columns on the punched card. After items had been coded and punched, they were entered into a computer subfile defined by bulletin type so that analyses could be made of bulletins either independently or according to channel.

The data was added cumulatively to the file and no decision was made at the outset concerning the size of the sample. Analysis of the first month's coverage provided clear indications of the major difference between bulletins and some idea of the level of inputs for different story categories. But it was not a reliable source for understanding the range of variation and the factors involved. Only after a further two months did the data seem to provide a good base for generalisations of this kind. The computer analysis therefore covers a three-month period (1 January to 31 March 1975).

The completeness of the record of three months' news avoided the many problems associated with sampling in a population whose degree of differentiation and variation were not fully known. The problems of sampling seemed to be particularly hazardous in the context of news stories, their development through time and their interconnections. Although the results show that many bulletin characteristics are highly predictable – indicating that a reliable sampling method could be devised – there was a further important reason for taking a complete record. The data provided a base not only for the numerical analysis by computer but also comparative linguistic and visual analysis and case studies for which the continuous and complete record was an essential requirement. Despite the intention to record the total news output a small percentage of bulletins were lost through operator error or machine failure. Those lost completely and therefore excluded from the statistical analysis accounted for 5·7 per cent of the total output time. Rather fewer bulletins were lost in January than in February and March but the proportion lost from each scheduled time and channel hardly varied. In other respects however, these bulletins are not representative of the output as a whole because a high proportion of them (54 per cent) were weekend bulletins, which as a rule form only 20 per cent or less of total news output.

September to December 1974

A series of meetings were initially held at which the basics of the logging sheet had been examined. In September, therefore, we were able to begin serious work on refining the sheet, the arguments as to its essential validity having been considered and alternative modes of operation rejected.

By the third week of November the television service of the university had implemented the technical decisions made over the summer. The machinery was in place and we felt we could try and record all bulletins and current affairs shows in that week.

Each team member took one day as their responsibility. This involved them in, first, checking the new video-cassette to be used, for a number simply broke on insertion into the VCR, and checking the settings of VCR and television receiver. The television receivers/monitors were permanently tuned to each channel but a malfunction on one outfit (VCR/receiver) could mean that the setting had been altered by a previous operator to cope with the problem. We found out that the automatic clock built into the VCR (and a prime reason for buying them) was not reliable enough for our purposes. So this meant that from midday to midnight the logger was on duty. While recording the bulletin (i.e. on first viewing) it was found that many columns on the initial form could be filled out, notably the headline and personnel items. But durations, categorisations and almost all the technical columns could only be completed by re-playing the VCR and reviewing the news. This meant that the filling out of the initial form took the fastest of us at least twice the real time duration of the bulletins.

After the first dry run the form was amended, notably in that the categorisation systems for interviewees and graphics were established. At this stage other subcategorisation systems for various aspects of the technics were discussed but it was decided that increasing the number of subsidiary codes that had to be learned would dramatically increase the possibility of error. So only the subsystems for interviewees and graphics were finally used.

A further dry run of a week was held in the first week of December incorporating the amendments to the initial sheet. At the end of this we felt reasonably confident that the sheet would work in the sense that we found in practice we could put the data onto it without too much difficulty. At this stage a rota was drawn up assigning each team member a specific day. Unfortunately this pattern had to be amended a number of times during the second phase of the project, notably because of other commitments especially at week-ends. As a result many of our worst human errors in failing to record specific bulletins occurred.

The two dry runs were also used to establish a system to check the logs. This was done by holding a meeting of the team to cross-check each logging sheet. Here differences in itemisation and story categorisation from one bulletin to another and from one day to another on running stories became clear. Decisions were reached collectively on how such penumbral cases should be categorised and this occasionally required a reviewing of disputed material. We also checked the total duration of the bulletins.

It was decided that this process of cross-checking should lag behind the recording process by a fortnight. In other words we would leave all bulletins in any one week for checking collectively a fortnight after they were recorded. The main purpose here was to ensure that we did not fail to transfer items which, although initially did not appear to be of interest, subsequently acquired industrial angles.

The team were by and large not specialists in the sociology of mass communications and therefore during this period we held an extensive series of seminars to review the literature in the area. This included inviting a number of scholars and other interested parties to give seminars on their own work and advise on the immediate problems the project was facing.

We had determined that it would be highly desirable to conduct observational studies inside the newsrooms and in-depth interviews with those concerned with industrial coverage, from both sides of industry. This work had become more pressing since we had adopted the strategy of following newsroom practices in the design of the initial logging sheet, so one member of the team was left in London to undertake this work. At one time this was envisaged as a major part of the project, but after our initial access problems described below, it became a crucial supplementary to the whole of the rest of the work. Therefore, although we believed that the examination of the output on the screen was the most important area, we acknowledged that any such examination could only be undertaken in the light of an actual understanding of the technical constraints and routine practices at work in the newsrooms. Unfortunately although some participant observational work has been done in British newsrooms very little has yet been published.

The project also decided to compare the choice of stories used by the television with another range of news material which was available to the production staff on a day-to-day basis. We decided that in addition to establishing an archive of the press which would include fringe papers we would also take the general wire service of the Press Association. This is an abstract of the full Press Association wire services running some four minutes behind it. It contains, for the benefit of smaller regional newspapers and television stations, a range of stories, foreign and domestic. We made no assumptions as to the balance and impartiality of this service but rather simply assumed that it represented a base of journalistic activity against which we could see the work of the television bulletins. The GPO and the Press Association arranged that a small teleprinter would be installed and working by the first of the year.

At this stage we also circulated to a number of unions, the CBI and the TUC a request for press releases and other publications as and when they became available. Although the response here was disappointing it was, as we found subsequently, an accurate reflection of the comparative unimportance of such formal inputs into the press in this area.

The second phase – January–June 1975

Preparing for observational work

Our relations with the broadcasting institutions were a matter of much concern to us. It has been true that at least until very recently, broadcasters have viewed the work of sociologists with at best, disdain, but in the offering of facilities for work, often downright hostility. We had decided that the focus of our research was to be the transmitted bulletin. While other research has concerned itself with the effects, professional practices and organisational structures of television, very little systematic analysis of the transmitted product has been undertaken.

Even though the task we set ourselves was to absorb most of our resources – some observation of the news-making process was assumed to be necessary, if only to sensitise us to the language of television. We wanted to have as much understanding as possible, in the time available to us, of the process of selection and production which underpins the news bulletins.

Our Research Director was an experienced television producer and through him we soon knew the technical repertoire of television, and the limiting factors associated with the use of film, video and so on. However, his experience was in current affairs departments. Journalistic practice and customary use of resources varies throughout television departments so we applied for observation facilities in the newsrooms of BBC and ITN.

We had heard that the BBC were hostile to our research and had even obliquely threatened us with the possibility of a copyright action. Before we approached the institutions the BBC had communicated with the project via the Principal of the university. After further prolix negotiations with the BBC we were given leave for one researcher to observe the newsroom and services for two weeks and the provision of certain other facilities, such as transcripts, was agreed. However, this fruitful outcome does not really reflect the ongoing relationship with the Corporation, which has publicly expressed itself to us on a number of occasions more or less hostilely. Nevertheless, senior personnel in BBC Scotland have been throughout most helpful although never concealing their central distrust of the work.

This ambivalence is best seen by contrasting the fact that at a local level eventually all team members were given brief access while at a national level the SSRC has been pressured by the broadcasters to limit the freedom of researchers financed by them by requiring them to sign contracts with the broadcast institutions before being given access.

Our relationship with ITN has been very much more straightforward. There has been no hostility and equally almost no cooperation. In such a situation it seemed to us legitimate to utilise, if you will, journalistic practices against the journalists. Despite the absolute refusal of ITN to allow our researcher to spend any time officially inside the newsroom except a walk-through visit culminating in watching the transmission of *News at Ten*, and an interview with the editor of ITN, she nevertheless managed to snatch five days' observation at ITN by well planned visits to journalist friends who worked there. Scottish Television (STV), the local commercial station, at the outset of the work sought an interview with us but after it had been explained that most of our focus was on national output and that the accident, as it were, of being in a Scottish university was not going to be reflected in a concomitant close analysis of STV's output, the relationship basically lapsed, although visits to local television newsrooms included STV's.

This description of the relationship of the project to various broadcasting institutions does not reveal the close association on a personal level with many of those responsible for the output, especially the current affairs output. This level of contact has permitted the project to remain au fait with the discussion about broadcasting as it has emerged around the Annan Committee of

Inquiry as well as being in touch with the development of broadcasting as an industry during the period of the study. Obviously this background, which has informed our work, has been largely concerned with the problems and thinking of broadcasting's middle management rather than senior policy-making officials.

At the same time as this work was going forward plans were made to interview in depth as many general secretaries of unions, business men, press departments and officials of the TUC and CBI as possible. This work was only partly carried out during the second phase of the project.

A major revision

The second phase of the project was to record the news and current affairs output on industrial affairs for the whole of 1975. The initial proposal envisaged an archive of some 240 hours of material. In the event this proved inadequate to cover the extremely heavy amount of coverage of industrial affairs in the first part of the year. By June the cassettes provided for in the initial scheme were full. Even with the buying of further softwear it had become increasingly clear that we could not continue to the end of the year. Not only would this have meant that the expense of the project would have been dramatically increased but also the work would have become more and more difficult to organise. The work of recording proved by repetitiveness to be extremely arduous.

Two further considerations became clear by the end of March. First the hypothesis had been generated that the bulletins were extremely formal with predictable patterns across the whole range of the data being observed in the technics column of the initial logging sheet. Second, the material was accumulating far faster than had been anticipated, despite the comparative lack of industrial coverage in current affairs programming.

All this was connected with the need to edit down the material. The project took a very broad view of what was of interest to it since it was felt it was better to keep material of marginal interest than to lose it irrevocably. Thus, in addition to industrial stories, most economic and business items were also being transferred. In addition, and increasingly important as the year progressed, the whole of the Referendum Campaign was being transferred. Aside from this broad interpretation of interest, for the essential work on the presentation of the news all slips and errors were being kept, as well as stories of particular interest for other reasons – such as the fall of Saigon. Further we knew we would have to keep one complete month of material for various linguistic and visual comparative purposes.

Thus it became clear that more material would be recorded than could ever be analysed in the third phase of the project. Indeed that the planned duration of the third phase – nine months – was grossly inadequate; and that the nature of the recording work was so heavy that many aspects of the analysis which it was hoped to accomplish before the start of the third phase would not be done. This last was an increasing source of frustration during the spring. The essential work on the linguistic and visual analysis was being prevented by the day-to-day routine of the project.

For all these reasons Referendum Day, 5 June, became an end date to the recording period and 31 March an end date to the observing of the repetitive technical data. Further, by stopping the work of computer coding and processing on the grounds that continuing it would yield nothing but further confirmation of the patterns already by that time being noted, it was possible to continue recording across the whole range of the output for another nine weeks. Without the decision not to fill out the second half of the logging sheet on 31 March it is unlikely we could have continued until June.

During this later period all items, duration, and personnel were noted as usual. In addition, further work, especially on the comparison with government figures to analyse the relationship between industrial activity as measured by the Department of Employment, could go ahead, and the interviews with the trade unionists and others could be undertaken. In addition towards the end of the Glasgow dustcart drivers' strike the team was able, with the help of the National Film School, to record onto video-tape their own brief reportage of the strike again for comparative purposes.

Thus the recording period was curtailed. The resultant archive consisted of some 260 hours of material containing all industrial and other news of interest; all other material in the bulletins (hesitations, errors, etc. – this last archive known as 'J' for Jokes inside the project although most of its material was not of humorous nature); video-cassettes for all the bulletins in May for comparative purposes; a month's archive of local Scottish news magazine programmes (which included the BBC's *Nationwide*) was kept; all weekly current affairs programmes and other specials were monitored and where appropriate kept throughout the whole period of recording; and finally, other programmes of interest to the project such as those dealing with the media themselves were also kept.

The project had also established a matching print archive for the whole of this period including the tapes of the Press Association wire service.

By mid-summer some initial comparisons between the output and the DE statistics and a few days' comparison with the Press Association had been done. The observational and interview work had been completed. This later became a rather more extensive task than had been envisaged. During this phase interviews were sought with as broad a range of trade unions as possible. Within this general range we selected unions that had received news coverage within the 22-week period of our study. For comparative purposes, we included some unions which do not typically receive extensive coverage. A series of interviews were held with the general secretary, a nominated deputy or the press officer of the relevant unions.

The basic pattern of statistics had begun to emerge from the computer. In addition preliminary case studies on two important running stories were prepared – the situation in British Leyland and the Glasgow dustcart drivers' strike, as well as some work being done on the hospital consultants' dispute.

The visit of the Committee on the Future of Broadcasting (the Annan Committee) to Scotland, which was planned for early in July, forced the team to somewhat hurriedly gather all this material together and present a short interim report on our findings.[5]

The ending of the recording period and the interim report to the Annan Committee marked the end of the second phase of the project. Hanging over the summer period were a number of problems mainly of logistics. Paramount among these were the obtaining of transcripts of the whole archive. The BBC had sold us transcriptions of their bulletins but these were in the wrong form for our purposes and anyway obviously did not cover the ITN output. It proved impossible to have the transcription done in Glasgow and therefore the work was done in London by a firm specialising in conference transcription work. Nevertheless it was not until October that the task was finally completed. In addition, copying the transcripts for each team member, copying the initial log, rectifying and completing the archive index cards lasted well into the third phase of the project. Thus the frustrations outlined above continued very much longer than was anticipated and ate into the third phase of the project.

The third phase

Aside from these residual organisational problems the work during the third phase had become largely predetermined by the pattern of work and interests of the team during the second phase. Therefore, the strategy adopted for the final phase of analysis was to tie the work to the production of these volumes; thus the work could be divided into two distinct areas – topics on which most work had been completed or at least begun and topics which had not yet been seriously tackled.

The topics on which much work had been done during the first two phases and over the summer included the survey of the literature of mass communication sociology especially as applied to the news; the observational material from the newsrooms; the computer analysis of the pattern of the bulletins; the relationship between television and the industrial world especially the comparisons with the DE figures; the reporting of industry by different media; the interviews with trade unions and some case studies. These are the contents of this volume.

During this phase our relationship with the broadcasters further deteriorated. We could not, indeed did not, know when we embarked on the task that media professionals would be so wedded to the notion that they were culturally fair, that any instances of significant absence or cultural skewedness that we could demonstrate would be received with hostility. Yet by December of 1975 some of our findings themselves were deemed to be so extraordinary that they were newsworthy. In a local Scottish BBC television news programme in which one of our researchers participated, the interview was proceeded by the introduction that we were to hear from a group of Glasgow researchers who 'allege' that the industrial news is biased. After more than the usual number of interruptions, the interviewer, an experienced professional, and apparently exasperated, said accusingly, 'But at the press conference this afternoon you claimed that during the course of the Glasgow dustcart drivers' strike, no striker was interviewed. Yet I myself interviewed a strike leader in this very studio.'[6]

Our researcher painstakingly reminded the interviewer that we were discussing national not local, news coverage. In fact as the results in this volume reveal, the strikers were not included in any of the twenty-one interviews shown on the national news during the thirteen weeks of the dispute. Such is the typical defensive professional response that a lecture at the local technical college became 'a press conference', and that despite attending the lecture and working for years on local television there was an apparent failure to distinguish between local and national news.

On the national level the weekly news and current affairs meeting inside the BBC on 21 November decided that a 'detailed and reasoned response' to the Project's evidence to the Committee on the Future of Broadcasting be made. The task fell to the then Chief Assistant to the Editor of Television News, Gerard Slessinger. In a somewhat jejeune document he sought to invalidate our evidence.[7] Normally, although flattered by this attention, the academic poverty of the 'rebuttal' would not have drawn forth a response. As it was, the manner of its distribution to third parties meant we needed to reply,[8] and a measure of public discussion followed.

All this is all the more remarkable given the public attitude to academic research on the part of the BBC. Sir Charles Curran, the Director General of the BBC:

> The case for unfettered research into the wide social issues raised by broadcasting needs no arguing. The BBC has always been wholly in its favour and it looks forward to the eventual increase in our understanding of the process of mass communication which must be expected. Our view is that research of this kind is best carried out in an academic institution. It is only in universities and other establishments of advanced learning that long-term projects of this kind can properly be undertaken.[9]

Aside from these diversions serious work was at last possible on the basic problems of visual and linguistic analysis as well as the development of further case studies. Since these are in essence the substance of the second volume of the study descriptions of the methodologies employed will follow, where appropriate, there.

3 INSIDE THE TELEVISION NEWSROOM

> If one of you were to go into a BBC newsroom and start talking to the journalists about their editorial responsibility, they might well scent pretension and shy away.
>
> Desmond Taylor, Editor, BBC News and Current Affairs

If the BBC, and by emulation the IBA, are to claim as part of their mandate to produce television programmes that they are performing a public service, then it is our opinion that the public must be provided with means to demand accountability from those institutions, and that this should include the possibility of access by academic researchers. Constitutionally at the moment the advisory boards form the only partial means of external inspection. Of these, Stuart Hood has said:

> These councils are the gesture the organizations make to accountability to the community. Like the Board of Governors they are chosen from a restricted section of the community in terms of social provenance, interests and age. It is the broadcasting organizations themselves who pick the members to serve on the councils, they have no representatives, no constituencies, and no responsibilities; it follows that their influence is minimal.[1]

We were given a measure of access as has been described above, but what follows in no way pretends to be an exhaustive description of the beliefs and practices of television news producers. Rather it is the record of a reconnaissance into alien territory. We used the results of observation to check on our developing hypotheses and recording techniques. It proved a valuable caution on the kinds of inferences we could make.

For instance, we had initially thought that the deployment of resources in the newsroom would give us some clues as to the importance attributed by newsmakers to particular events. Our observation soon made us aware of the complexities governing the use of all resources. Satellite inputs are, for example, as much to do with the BBC's accountancy systems as to the newsworthiness of the events they carry. The BBC were able, for instance, to give lengthy coverage of interviews with John Stonehouse and his family in Australia on the tail of a satellite transmission originally booked to cover the England–Australia test series. The use of satellite material was thus responsible for these 'scoops'.

One other use of the observation was to reassure us of our choice of story categories for documenting the bulletin output. We had experimented with several category systems and finally decided on the simple journalistic one of relating stories to the correspondent or 'desk', foreign, economic, etc., which would be responsible for their development. Had we chosen a system which did not parallel news processing we might well have lost sight of those moments of decision in the newsroom when a story initially developed by one set of specialists, moved to another 'desk'.

Industrial events, for instance, penetrate as news stories initially via the industrial newsdesk, but they can sometimes become part of an economics package. One very important effect of newsroom organisation and the processing of the news, is to compartmentalise events as they arrive in the newsroom. The decision to shift a story from one compartment to another, that is from one newsdesk to another, demonstrates the new contextual relevance accorded to that story. These decisions are not explicable as the exercise of journalistic practice only. They are more importantly decisions that make for a particular interpretation of events and they provide the basis of the interpretive framework for the raw events.

It became clear through this and other information gathered during observation that the compilation of a bulletin was not dictated solely by journalistic news values or a sense of public service. The technology of television news production itself helps to determine the shape and content of bulletins to an extent which the newsroom personnel seem reluctant to admit. Indeed, what Galtung and Ruge have to say about the function of cultural factors in the selection of foreign news could well be extended to include technological factors as selection determinants.[2]

During the period of observation, we had many informal interviews with workers in the newsrooms. With few exceptions they were at pains to assure us that the news was undistorted and uncensored and was governed by journalistic criteria and expertise. Even in the brief time available for observation it became obvious that working under the constraints and pressures endemic in any news-processing enterprise, the participants themselves become unaware of the many factors including the obvious cultural ones that govern and shape their output.

The researcher primarily responsible for observation in the newsrooms presented a series of papers to the research team, all of whom themselves at one time or another, at local or national level, visited television newsrooms. The papers became known as 'Penelope's Diary', a reflection on the necessarily impressionistic background they afforded, any serious attempt at rigour being precluded by the minimal access given us. We include them here.

The newsroom at Independent Television News

The newsroom at ITN is on the first floor of a large, architecturally anonymous building in the garment district of London, north of Soho. Originally planned as an office block it was bought by ITN half-way through the building and the result is a partially custom-built studio and facilities. The haphazard design has repercussions in the day-to-day processing of news, as access to the control room from the newsroom is quite difficult and appears to contribute to pre-bulletin tensions.

The newsroom itself is about 3,000 square feet (60 by 50 feet) packed with clusters of desks and there is little room to move. All around the central square are small facility rooms, four small film-cutting or viewing rooms (about the size of our own video room),[3] one larger room housing the agency input machines and reference books. On one side are the correspondents' rooms, too small for all correspondents to use at one time, and the assistant editor's room. The producers also have a room but it is upstairs and too far away from the locus of their activity to be used often. People work in close proximity. The clusters of desks are little more than three feet apart and until I began to recognise particular faces and working relationships, everyone seemed equally anonymous, casually dressed and informal in their behaviour. The newsroom is certainly no place for reflection, even if there were time.

In the middle of the room among the groups of desks is the assignments desk, the liaison point between film crews, reporters and processers. One end of the room is occupied by the agency input section from which news agency stories and pictures are constantly printed out. The copytaster has a desk at the entrance to this facility. He acts as a filter for the newsroom. He selects the stories, knowing the intended shape of the bulletins and passes them on to the sub-editors responsible for that story. Only if some unanticipated event or unusual development in a news item occurs, will he go directly to the producer. The copytaster is clearly entrusted with the dissemination of information based on accepted journalistic criteria. He constitutes one of the 'gatekeepers' which help to ensure that the news is always – in Boorstin's phrase – 'the olds'.

Several times I heard ITN journalists call their company 'the lean cat', in terms of facilities, in contrast to 'the fat cat' BBC. The image ITN has of itself seems to be that of a spare but flexible organisation, capable of quicker draws on the gun than the old-fashioned, bureaucratic 'Beeb'. One senior journalist told me, 'I couldn't get a job at the BBC, I'm not slow enough.'

The staff on the whole are very partisan about their news bulletins compared to those of the BBC. They talk of the opposition as wooden, over-formal and dull, and pride themselves on their more conversational style, better pictures, pace and immediacy. Some, however, especially those with prestigious Fleet Street experience, have reservations about the 'sloppiness' of ITN's journalism. There is very clearly some unresolved tension between the craft of newspaper news writing and that which has become the style of television news.

The daily routine

The day begins at 10.30 a.m. for the editor and staff on duty. There is always at least one duty sub-editor arriving at 6 p.m. and working until midnight and of course there are typists and other secretarial workers on a normal office time-table. Editorial staff work a complicated but routine shift system, sub-editors and producers have to be prepared to adjust their working schedules to the volume of news and the availability of production staff.

Basically, the *First Report* team (midday bulletin) work a four-day week, unlike the Early Bulletin and *News at Ten*, where there is a degree of swapping between producers and chief sub-editors. The producer, chief sub-editors and some of the sub-editors are permanently assigned to *First Report*. This is so because *First Report* is seen as serving a different function from the Early Bulletin and *News at Ten*. *First Report* is seen as a prestige news programme, despite the midday slot. It is seen by the staff of the other bulletins as something of a public relations exercise, directed at the Independent Broadcasting Authority and the government. In the absence of a current affairs department it also serves the function of showing ITN's journalistic stability by affording the possibility of extended interviews to politicians and other people in the news, whose goodwill can then be exploited in the later bulletins. One editor said, 'Robert Kee [the presenter of *First Report*] can do his Robin Day aggressive bit when there are only housewives to watch.' The producer of *First Report* enjoys a status similar to that of the two assistant editors to the editor of ITN.

However, apart from *First Report* interviews, everything gives way to *News at Ten* – all facilities are primarily directed at producing ITN's best at 10 p.m. There are two senior producers for *News at Ten*, the junior producers for the Early Bulletin 'act up' or deputise for them when necessary. They, in turn, are deputised from the ranks of the four chief sub-editors (one of whom hates producing bulletins). He says it is not necessarily a journalist's job, so he, to square the ranking system, is called Chief Writer and is only asked to produce in emergencies. He said, 'If I'd wanted to be a bloody producer, I'd have gone on the stage.' It is noticeable that ITN use more theatrical terms than the BBC, for ITN it's 'the Green Room' and for the BBC, 'the Hospitality Room'; for ITN 'the Bar' and 'Club' to the BBC;[4] 'Producer' – ITN, and 'Editor' – BBC; 'Scriptwriter' – ITN, and 'Sub-editor' for the BBC.

Producing *News at Ten* is the plum news-processing job (as distinct from news gathering) apart from the three editors and Ryan (ITN's Editor-in-Chief) himself. Early Bulletin teams work two to three weekdays and alternate weekends, and *News at Ten* work two to three days on and two to three off, i.e. nine days in fourteen. The two senior producers have more control over their working schedules than those lower down. The job of producing a news bulletin to time is a heavy responsibility, especially for the producer and chief sub-editor, who have quite rigorous working schedules. One producer said that tension is an occupational hazard: 'One's nerves go and it's time to take a holiday.'

Although the shift schedules mean that the bulletins are seldom prepared by the same team, producers do have their own preferences and say that they produce their best bulletins and achieve their own characteristic style when they work with writers, reporters and the director of their choice. They see the personnel involved as influencing the bulletin. For instance, two producers told me that one director (the person responsible during transmission for the technical co-ordination of items) was less competent to handle complicated series of video, sound and graphics, so when this director was on duty they often simplified the running order, especially when rehearsal time was limited. The newsroom staff seemed to think that there is more differentiation in bulletin style than I think can be seen on the screen.

Directors, although an integral part of the news-processing team, are considered by journalists to be part of the technical staff. They are not considered by the journalists to be of much more than 'button pushing' importance to the bulletin. One journalist said 'they're just bloody mechanics, not paid to think'. There's a clear blue-/white-collar division in the newsroom of which journalists are the élite. However, directors do interfere with the semiotics of the news bulletins in many ways since they choose many of the stills used on colour separation overlay behind the newscasters and when time is short, during rehearsal, they suggest the cuts in film or tape. ITN have six to eight directors, some of whom are freelance. Everyone's favourite director is a woman who has been with ITN since it began. Although traditionally women can rise from studio manager to director, there does not seem to be much mobility for women in news processing generally.

Newscasters at ITN are from journalistic backgrounds and are members of ITN's National Union of Journalists Chapel. They often select stills and re-style the scripts they have to read. They also make suggestions as to word and picture 'match'. The two newscasters of *News at Ten* regard 'the bongs' (the headlines that coincide with Big Ben) as their own and influence both the writing and selection of accompanying visuals, though the sequence is determined by the producer. During my observation I was aware that some newscasters are more influential than others and in some cases were able to affect radically items in the bulletin.

There are preferences in the newsroom for each of the newscasters and they seem to illustrate the ambiguities that the ITN personnel share over every aspect of news production. While one faction support Sandy Gall or Andrew Gardner for qualities that seem very 'BBC' such as authority, continuity, unflappability, precise diction, etc., another will praise Bosanquet's humanity, his ability to ad-lib, rugged face, glamour of his private life, and so on. This ambivalence as to what constitutes a successful bulletin runs through most replies to the questions asked. Those replies were full of contradictions. On the one hand, the news was thought to be necessarily objective, authoritative, 'untouched by human hand'. On the other hand, it was seen as necessary that the news should hold a fickle and demanding audience by techniques which have to do more with 'show biz' than with the provision of information.

There is an editorial meeting at 10.30 a.m. each day. This is attended by ITN's Editor-in-Chief, the assistant editor responsible for news intake, the producer for the day and correspondents. One scriptwriter said of the meeting, 'I'd love to be at one of them.' He clearly attached much significance to the exclusivity of the membership. But he later said that the producer had a completely free hand in preparing the news and was never under editorial pressure. This is the major news policy meeting and governs the shape of the bulletins of the day. Likely running orders are sketched out and facilities deployed to cover events. These, of course, are subject to change throughout the day.

First Report has first call on facilities although Early Bulletin continues to prepare for the evening; but everyone leaves the narrow gangways clear for the *First Report* team. Surprisingly and contrary to expectations, there doesn't seem to be much reading of newspapers – the bins are full of almost new newspapers. Most scriptwriters, producers and reporters I spoke to, say they try to listen to the 8.00 a.m. news on the radio, which is of course already the result of a filtering process.

There is a meeting before *First Report* transmission at which all senior staff can sit in – but it is not obligatory. It depends on the volume of the news and the intended use of film or tape. *First Report* staff complain that they have little flexibility in writing and editing, since much of their material will be needed, maybe in a different form, by subsequent bulletins. The producers for the later bulletins prefer tape and film left raw – as far as possible – that is, unedited.

The Early Bulletin staff have a running order meeting at 5.00 p.m. At 7.00 p.m. another meeting is held for *News at Ten*. This meeting is a policy one and crucial – since it is here that justifications to the editor are made, and weight is brought to bear by the Editor and his assistants on the producers of *News at Ten*. This meeting is also attended by the next day's producers and functions as a 'look-ahead' meeting for the next day. It is therefore quite a public forum in which to be criticised or censured.

The 10.30 a.m. and 7.00 p.m. meetings are considered the crucial ones by the newsroom staff.

Weekends are leisurely. One typist regularly brings her 18-month-old baby in at weekends. The news at the weekend is traditionally thin and at ITN two of the six Ampex video machines are leased to London Weekend Television for their sports coverage. Video tape recorders are also leased throughout the week to commercial advertising companies and editorial staff complain regularly that ITN's slim facilities are being further impaired by the need to pay rent. ITN's second studio is not permanently leased to any single company. News programme requirements take priority over all commercial bookings and, in fact, both studios are used during general elections. At the weekends there is a heavier usage of agency film especially UPITN, a shared company rather like BBC's Visnews, which is reflected in our data.[5]

During the day adjustments are made to the running order, reflecting the volume and importance of news intake and the success of reporters, correspondents and incoming film. For instance, on 22 January 1975, the Political Editor was waiting all day to talk to the Prime Minister about the Cowley dispute. He phoned in, complaining that 'the BBC are all over the place', and he was having trouble. A place was reserved for the film in the running order – but a 'fail safe' item was prepared in which the newscaster would refer briefly to the Cowley event and the space would then be filled by a prepared human interest story. It is interesting that primacy or hardness of a story often does *not* lie in the event itself; it can be affected by the mode and success of the coverage.

From the beginning of each shift the editor allocates stories to sub-editors under a *catch-line*, by which it is known throughout the process. It then becomes the sub-editor's job to condense the story according to the time allocated to it by the producer, to chase up films and library stills and to view and supervise the editing of film or tape. If a story requires complicated processing – a radio tape from an outside or overseas reporter to lay over film, etc. – that's his or her responsibility too. The main decisions about the use of visual material will have been made at the meetings by the producers.

The sub-editor's job is to implement those decisions and to accommodate them to changes in time allocation. They do have quite a broad licence, however, on selection and override film editors during a conflict.

I have heard arguments in the film-cutting room about the selection of film but on the whole the scriptwriter wins – this is a common source of tension. A film editor would have to go through the producer if he wished to impose his judgment. (Significantly, film is referred to by the name of the reporter, not the cameraman.) One film editor insisted, during the Red Lion Square demonstration in September 1974, that the scriptwriter had chosen the least newsworthy piece of film. In the enquiry into Kevin Gately's death the untransmitted film revealed the incident in which he was carried off after being struck. Recently the Windsor Pop Festival episode when film was again subpoenaed, has made it obligatory for the producers on duty to watch all footage of film from ITN cameramen.[6] They have always been formally required to do this, but in practice shortage of time has usually meant that the scriptwriter was delegated to decide for himself.

Scripts are returned to the producer via the chief scriptwriter who is the workhorse for the producer. The producer begins, only about an hour before transmission, to work at the final order and time allocation. During this time, the chief scriptwriter is hecticly cutting or expanding scripts to suit the precise timing of the bulletins. The final bulletin can still be in the process of adjustment up to forty minutes before transmission. The director will have supervised the technical aspects of video, graphics, stills, etc., with the producer although these have been developing all day. Newscasters appear some while before transmission to check over their pieces for pronunciation, accent, timing, and to familiarise themselves with the content. Finally, even if there is only time for a ten-minute rehearsal, producer, director and personal assistants go to the control room and newscasters to the studio for a run-through, right up to fifteen seconds before transmission.

The frenzy of the last hour is amazing. I suspect that it is deliberately so, or at least that there is little attempt by the producers to minimise the countdown tension. (I know a lady singer who always runs back to her dressing room seconds before a curtain up because she wants to appear on stage with 'her juices flowing'.) Both producers I've spoken to have said the same thing: 'good rehearsal – bad show' or 'nothing like a panic for a good show'. One producer prides himself on keeping an atmosphere of 'sizzle' on his team, and himself plays the boozy journalist, always threatening to pass out. Of course he never does and is generally regarded as efficient.

Maybe it's true of all newsrooms, but ITN is full of jokes about 'being pissed' all day long, although in fact there wasn't much drunkenness. Obviously in work where long periods of boredom alternate with periods of hectic tension and where mistakes are very public, drink is an antidote to nerves. I suspect that the 'game' of drunkenness played at ITN has more to do with showing detachment and 'cool' during the process: the outcome of which is that no one is totally at ease. Perhaps some journalists find this kind of formular processing of their 'professional commodity' distasteful and a boozy insouciance helps them to accommodate it gracefully. It may sound like over-psychologising, but from 10.30 a.m. onward most of my enquiries were answered by 'I'll tell you when I'm pissed', or after 3.00 p.m., 'I'm too pissed to remember', and after *News at Ten*, 'Now let's go and really get pissed'.

The fact that popular mythology has it that ITN's Editor is supposed to be watching every bulletin from some dinner party might add to the sense of strain.

What is a 'hard fact' in the television newsroom?

Every person I spoke to, both processers and news-gatherers assured me that the news bulletin reflected the most important 'hard facts' available about events in the world relevant to ITN's audience. I received, in answer to a request for a definition of 'facts', the usual plethora of replies which add up to 'a fact is decreed by journalistic intuition'.

A 'hard fact' appears to be information delivered as being disclosed by an 'unimpeachable source'. The validity of sources is again the perquisite of journalistic intuition, it seems. Any form of conjecture or an investigative approach would 'soften' a fact. Since television newsrooms are prohibited by act and charter from comment, it follows – or, rather, it is taken to follow – that the news bulletins contain only 'hard facts'.

In practice, during my observation, 'hard facts' became harder or softer, more or less likely to be included in the bulletin, if they were accompanied by live or exclusive film. Exclusive interviews have a very high rating, especially when it is known that the rival BBC may well be after the same story. What is a 'hard fact' is determined in many ways. Shared journalistic norms ensure that there will be an overall similarity in selection between BBC and ITN. However, ITN staff say that despite the comparative paucity of their film resources, ITN uses better film than BBC. ITN film is thought to be more 'immediate' and have more 'pace' than that of the BBC.[7]

What dictates the primacy of 'hard facts' is more variable in television than in the press. There are more technical inputs to provide potential stories, i.e. film, video, stills, interviews and material provided by specialist correspondents and agency services. The successful acquisition of items through any of these resources, or combinations of them, alters the importance attached to the event itself by television teams. It seems that the norms by which these journalists rationalise their selection of news events are inherited from newspaper experience and are no longer adequate to explain the criteria which operate for the selection for television news.

News producers' views of audience

There seems to be a great deal of confusion here as to what kind of animal an audience is. Even though middle-class literary standards of clarity and style operate at a professional journalistic level, the people I've spoken to complain of what they see as the audience's low level of tolerance for information. This they say, requires that the news, like medicine, be made palatable. The tele-journalists' problem is that they cannot address themselves to a known partisanship as do newspaper journalists, so selection and presentation are directed at the 'middle-range'. One producer, a very deliberate cynic, says he has in mind the family of the Kellogg's Cornflakes commercial. However, at a later stage when I asked him whose praise he valued most for the bulletin he produced, he said, 'other producers' at ITN'.

There is clearly a more direct pressure on ITN to 'hold' its viewing figure, if for no other reason than that they have in *News at Ten* thirty minutes of network time which, with only one channel, the commercial companies must covet – especially if viewing figures fall. Viewing figures must be kept high to justify the double rates charge for advertisers for the slots before, during and after the *News at Ten*. Clearly professional journalists do not want to come to terms with the precise classifications used to describe 'consumer groups' by advertisers who depend on the 'pull' of *News at Ten* to justify the high advertising charges. So the middle range to which the ITN news is directed must be to some extent an inherited BBC middle-class option, tinged with a necessary 'popular' approach – itself partly the inheritance of ITN's early innovative days.

The BBC's emulation of ITN's techniques – for instance the use of two newscasters for their main bulletin during the period of the study – has nevertheless meant a convergence of presentational styles and content which is demonstrated below in this and the subsequent volume.

Where do ITN journalists go when they are old?

Unlike the BBC, ITN is a small organisation concentrated on the production of three news programmes. Apart from the secretarial and technical staff, there are about eighty NUJ members. While this allows great flexibility during the preparation of the news, every specialist and every skill is on hand and last-minute changes can be made without lengthy departmental negotiations, as appears to be the case at the BBC. It also means curtailed promotional possibilities.

For the internal processers, *News at Ten* is the last post. The two senior producers are between 40 and 50 years old, the juniors slightly younger. The correspondents' desks are staffed by young men. The industrial editor is in his early 30s. The promotional ladder for reporters, who aspire to the correspondents' desks, is virtually non-existent. Some go to the BBC, and there is some circulation among the regional companies especially by journalists who want a chance at 'in-depth' current affairs journalism. Others go into public relations work and a few to advertising, although this is considered a 'low-life' occupation by those I spoke to. This situation means that ITN recruits very few people, except from amongst skilled journalists. There is, in any case, no time at ITN to train anyone. At the moment they have one graduate trainee who is personal assistant to the Home Editor.

The fact that ITN's staff is a static ensemble contributes to a conservatism of style, and a reliance on trusted formulae. It is thought that there are moves to redesign *News at Ten* which may be prompted by pressures from the companies to regain 10.00 p.m. prime time for their own use, especially since the drop in viewing figures. There is one faction who want *News at*

Ten to emulate tabloid stories, à la *Daily Mirror*. 'Let the BBC give the news straight and give the audience something different on ITN.'

ITN's public relations activities include inviting various interested parties such as personnel of the commercial television companies to watch bulletin transmissions. Every night I was there parties were being entertained and given the guided tour – culminating in watching *News at Ten* from the control room.

It's becoming clear that the requirements of a highly centralised, mechanised news processing organisation which is tyrannised by the demand of constant, unbroken 'flow', increasingly encourages the development of news sources in which events, information and such are pre-packaged with the end requirements in mind. The more successfully events are presented to suit television requirements, the more chance they stand of being selected. I do not suggest that this alone explains why we get highly coded coverage, but it is likely that the ability of some sections of the community to anticipate and supply the needs of a conveyor-belt news process, helps to maintain 'news as a reinforcement of consensus'.

External pressures and constraints

I've spoken to three producers, three chief sub-directors, about six sub-editors and three PAs. They all insist that at no time have they received a directive from the editors about inclusions or exclusions from their bulletins. Some of them told of incidents – an interview with an IRA official, an interview with a politician which was considered too extreme – which were dropped as a result of the direct intervention of the editor. I think what most of them meant is that there is little formal process at ITN for issuing statements of house policy – though house-style books do exist. Obviously, both BBC and ITN rely on selected recruitment and conditioning of personnel to produce a continuing auto-censorship. At the BBC it is reinforced confidently by the management in written directives – not, however, accessible to all personnel.[8] At ITN hardly anything seems to be written down. Although the editor is remote, he is thought to be ever present and watching. There is a little overt directive – if the editor wants to censor or censure, he does so directly, and in that way each journalist at ITN is in a direct relationship to whatever pressures make the editor jump.

The BBC paradoxically, through its complex hierarchical departments, allows more deviance. Offenders can always be kicked upstairs to Current Affairs, or in severe cases of reprimand, shunted off to the provinces or obscure managerial positions.

Independent Television News is, on the other hand, a service to a commercial television network. Lady Plowden, the IBA's chairperson, speaking to advertisers in July of 1976 said 'Like other industries, Independent Broadcasting aims to thrive. It aims to make a profit . . .'. In this she was no more than reiterating a number of government reports from such sources as the Prices and Incomes Board (Report No. 165) and the Commons' Select Committee Report on the Authority itself. But acknowledging the commercial nature of the Network does not of itself demonstrate a connection with commercial pressure on the output, a pressure the Act and the Authority are specifically designed to avoid. (The advertisers, as witness the article in their trade paper *Campaign* in August of 1976, are not averse to letting their views in general on the output be known.)

Usually the producer or the chief sub-editor handle immediate phone complaints after every bulletin, and the general rule is to take the name and address only of those complaints where evidence of inaccuracies can be supplied, in which case the assistant editor replies. Recognised names or institutions are referred on directly to the most senior editor on duty.

Finally, my impression is that none of the people I've spoken to are unaware of the political and historical antecedents that form television journalism. They see themselves as doing the best they can within the limits imposed by the state, the ITV companies and the IBA. What they seem less aware of is how those constraints are reinforced in the techniques and processes of the production itself. The intensity of the activity imposed by fifteen- or twenty-five-minute bulletins leads to a day-to-day attitude of beating the clock, for its own sake, in the most 'exciting' way possible. The conventional embargo on investigative journalism and the definitions of impartiality imposed by the Act lead to extensive use of formulae and conventional news agenda setting.

Decisions as to the content and weight given to particular news items are made very close to the directorship of ITN, so beneath the producer most effort is concentrated on the development of skills that are directly related to the solution of technical and time problems. News gatherers seem fairly free to explore their subjects, but they know that whatever they submit will be subjected to the processes and requirements of the bulletin. The processers are, in turn, only allowed a small range of competence and judgment – an ability to compress large amounts of material and an understanding of the technical requirements of the medium. Perhaps that's why tele-journalists (particularly the processers – all of whom have had newspaper journalistic experience and presumably have imbibed some of the ideology of 'the fourth estate'), when faced with the particular anonymity of team journalism required by television, look for excitement and satisfaction in the immediate showbizziness of their job. For some the hope is to write a bestseller which will release them from the routine of the newsroom. Two journalists I spoke to were writing plays and novels (not an unusual activity for professional writers) but they were writing them in the hope of getting a 'bestseller' on which to retire from television.

Inside the BBC newsroom

The first impression a visitor receives of the newsroom at the BBC is its isolation from the rest of the Television Centre and other activities of the organisation. Television Centre is shaped rather like a polo-mint, the circular corridors serving tangents of studios, facilities rooms and offices. The Spur, where the newsroom is based, is exactly what the name implies – a hook off the main building. The news operation is referred to as 'the Spur' – workers there referred to themselves as 'we at the Spur'. This is rich in implication as well as actuality. The workers in the Spur prefer to see themselves as engaged in a separate and possibly superior activity, contiguous to, but not embedded in, an enterprise concerned with the production of entertainment.

The Spur houses not only the newsroom, the centre of operations, but also film and video processing departments, an information library, a small interviewing studio and the offices of the senior editors and specialist correspondents' rooms. It has its own instant food bar, but no drinks bar. However, even when people from the Spur use the BBC Club, there is a tendency for them to use the small bar rather than the main room of the Club, possibly an attempt to establish an 'El Vino's' in what otherwise are anonymous surroundings.

Unlike the newsroom at ITN, Television Centre is a long way from any cosy restaurants or pubs. People do not, generally, go out in their breaks but use either the Spur facilities or the main canteen – which provide the kind of utility food to be found in most canteens. So unusual is 'eating out' that it is rumoured that parties making for the restaurants of Shepherds Bush herald some kind of palace revolution in the newsroom.

The atmosphere, in comparison with ITN's newsroom, is sober and formal and certainly much quieter. Whereas at ITN there seemed a deliberate 'back stage' pandemonium – every bulletin a first night – here at the BBC there is a feeling of urgency and activity well under control, a deliberate sense of every man keeping his head down in the trenches. Everything – dress, deportment, furnishing and decor – speaks of decorum and order.

The newsroom itself reinforces the sense of being sealed off from the world. Although on two sides there are large windows looking out over White City Stadium and the development chaos of Willesden, the newsroom appears to be heavily sound-proofed to facilitate loudspeaker announcements from Broadcasting House and the endless reports from radio and film crews at the lines-input control in the newsroom. The BBC newsroom is clearly much more luxurious than ITN; there is more space, more television monitors, more service facilities; there are dark coloured carpets and good matching furniture – nowhere that 'improvised' look of ITN.

The newsroom here is undoubtedly better endowed with facilities. First there are the obvious advantages of being part of an organisation with a traditional and respected place in the network of international newsmongering (when BBC correspondent in Washington, Charles Wheeler was said to enjoy a status close to that of informal ambassador); second, the BBC as a continuous and centralised institution has many lines of access to those other institutions which are likely to provide the mainstock of news – the chief example being the government and civil services.

Last, the immediate physical facilities appear to be superior to those of ITN. There are more staff, more rooms, more everything I am told. However, this does not necessarily mean that those facilities enable the newsroom to function more efficiently. Already I have spoken to several junior newsmen (sub-editors) who complain of the bureaucratic delays that accompany the advantages afforded by being part of the Corporation.

The obvious difference here in the organisation is that the intake section dominates the newsroom. This desk is immediately by the newsroom entrance and leading away from it is the department which prepares 'futures' (anticipating events coverage) and features. Here the activities of film camera crews, reporters and the BBC regional studios are co-ordinated with the requirements of the developing bulletin. Whatever hysteria there is emanates from here and it is functional hysteria – not the 'showbiz' displacement hysteria which appears to be the case at ITN. Again the military analogy came to mind – this desk is the centre of operations, complete with wall charts depicting the place, time and arrival of camera crews and reporters deployed on news-gathering missions.

Day-to-day routine in the newsroom

The news for both channels 1 and 2 is produced from the same newsroom using the same staff (except for newsreaders/presenters) on a shift basis which varies for different grades but normally means a two-day production schedule, with an extra day's planning for senior duty editors who are the BBC equivalent of ITN's producers.

The daily routine of news processing is pretty much the same as that at ITN – in fact ITN adopted, along with personnel, the BBC routines of television news production. They altered the presentation and to some extent, the content, but the production system remained a slimmer, simpler replica. The replication of programme profiles is obviously importantly determined in this shared method.[9]

One team under the control of the editor of the day produces the 5.45 p.m. bulletin and *News Extra*; another team – the *Midday News* and the *Nine O'Clock* bulletin. Each team is responsible to one editor of the day who in turn, is responsible to the editor of Television News. Beneath editor of the day is a senior duty editor who is the right-hand man responsible for the allocation of stories to sub-editors and for their development throughout the day. Each process – film, script and specialist correspondent's pieces – are filtered back through the senior duty editor who constantly tailors them to the shape required by the editor. At the SDE's side sits the shift leader who has to be a crack typist and a cool head (always a woman) who constantly types and times the scripts as they are produced. All scripts are stencilled and distributed to all personnel and the newsreader, who sits throughout the process absorbing the information, checking on pronunciation in preparation for his/her performance on the bulletin. As at ITN, the sub-editor is responsible for accompaniments in terms of film, video, stills and graphics for his story. He may, of course, be delegated to work with a correspondent on a 'package' story but everything, no matter from what source, goes through the SDE. I have a feeling that the visuals for BBC news bulletins were more carefully scrutinised by the editor of the day than at ITN. In fact the whole operation seems much more authoritatively controlled, with more reverence in every way, for seniority and status. The newsroom starts at 9.30 a.m. with an editorial conference. Both intake and output editors are present. Material which has been prepared by the 'futures' and 'events' departments is inspected by the two editors for the day, home and futures editors and the foreign editor. Very little except the pre-prepared diary items is considered at this conference. Broadcasting House News Conference (Radio) takes place concurrently and there is a direct link between the conferences.

Ultimate decisions on bulletin items rest with the editors of the day. There follows a round table and the allocations of stories to be researched, written up and matched with film and other visual material. Each team consists of editor of the day, his workhorse the senior duty editor, eight sub-editors and scriptwriters and the shift leader. These then call on the personnel and facilities of: the input section (futures, events, incoming agency material), film and video section, the graphics department, the specialist correspondents and, of course, the reporters who act as producers in the field to the camera crews who accompany then in producing on the spot reports.

The midday bulletins begin to take shape at about 11.00 a.m.; there is not usually much film on the midday news. It is not difficult to assemble the film but it is something of a nuisance to prepare (edit, etc.) and since at the BBC a full rehearsal is obligatory, the midday news is kept as simple as possible.

The patterns of preparations, meetings and full rehearsals, remain similar throughout the day for each of the bulletins.

The 'toddler's truce' – those periods of the day when children are thought to be watching – affect the use of films and some reports. Subjects thought to be offensive for young viewers might be held back from the 5.45 p.m. news to appear later in the *Nine O'Clock News*.

The *Nine O'Clock News* is clearly the most important bulletin of the day. All facilities are deployed to give the BBC's best at 9.00 p.m.

The day team producing the bulletin is particularly interested every day to see how well ITN do by comparison one hour later. Both ITN and BBC view each other's bulletins in their respective newsrooms. If BBC is to 'scoop' ITN it is more likely to be at the prestigious evening bulletin than at any other time, although there have been times when the hour's difference has been an advantage to ITN. This week, for instance, we have been watching the impending fall (liberation?) of Saigon. The difficulty of getting film and on-the-spot reports of co-ordinating air freight and so on, has meant that getting the news from Saigon has been a continuous cliff-hanger in the newsroom, strengthened by the fear that ITN with an hour to spare would scoop them. Often there are chilling examples of journalistic objectivity. Pieces of film or video showing carnage and mayhem are said to be 'fabulous pieces of film' with the added 'pity we can't use them'.

A full rehearsal of each bulletin is considered obligatory at BBC and is only departed from in exceptional circumstances such as the day of the Budget, when the newsroom facilities were stretched. The division in the control room between news production and technical production is very obvious. The editor of the day is in charge throughout and produces the programme, making last-minute cuts and switches often during the transmission. There is a floor manager in the studio giving, where necessary, visual cues to the newsreader, cameramen and graphics assistants.

Only two of BBC's newsreaders are journalists. Until recently there has been a tendency in the news department to have Equity rather than NUJ members as the newsreaders. This may well be because it is assumed that only actors could create the necessary anonymity required for neutral news presentation.[10]

Correspondents

BBC have traditionally given seniority to the foreign desk, and a sub-newsroom organisation exists for reasons of intake and co-ordination, to support the flow of foreign news. This status also reflects BBC's important international reputation during the last World War. Of the reporters I spoke to, most felt that in terms of life style the foreign desk was to be desired. However, others thought that journalistically the political desk afforded most scope. None of them wanted to be an industrial correspondent when they grew up and liked least of all working as a reporter for this desk. 'Nobody likes being a runner for the industrial correspondent.' 'Who wants to go to boring TUC press conferences or Barrow-in-Furness?' They complained of the 'caginess' of trade union interviewees, the routine replies they were likely to receive from them, and the whole cautious approach of industrial representatives, both TUC and CBI, to interview situations.

Film crews

The same blue-/white-collar division exists here as at ITN. Although 'star' cameramen are given due praise within the newsroom, their activities are directed by the reporter or correspondent who accompanies them in the field. When they are unaccompanied they are given precise indications as to the nature of the coverage required from the editors in charge. (The sub-editors seem to envy the reporters their role as producers in the field. Some think that this is a responsibility that should be left in the hands of the processing department. The fact that ITN scooped them on the Cyprus coverage was attributed by one sub-editor to lack of skilled co-ordination in the field. It is considered quite a 'perk' to be given the task of directing coverage of events such as the Paris air show.)

The news in terms of visuals is prestructured well in advance. Looking through the camera crews' logbook I can see how, for instance, the intended film of a 'show of hands' vote at the London dock (Friday 4 April 1975) was anticipated. Scaffolds were hired and placed in the morning of the vote. A particular window was staked out in the yard overlooking the meeting place. The cameraman in charge of the film coverage told me that he would be looking for 'action' and that 'panning' shots of the raised hands had been requested as a priority. When I asked him whether he felt that this might misrepresent the mood of the meeting, he said that the purpose of film was to liven the news and that the criteria of this film coverage, was action. 'It's up to the reporter to get the story straight.' Such routine beliefs about the status of film in news reports seemed to evidence little awareness of the powerful mediating effect of this tight organisation of shots and positioning of the cameras. 'We record actuality,' remarked the same cameraman. In the event the focus of coverage allegedly falsified the containerisation dispute in the London docks. Halloran et al. gave a very good account of how this preplanning can affect the subsequent coverage of an event.[11]

Interviews

There is obviously some tension in the relationship between television journalists and their intended interviewees. Who is doing what and for whom, seems to be the problem. Are the interviewees sought after as a necessary part of bulletin content or is the news being used as free PR operations for different groups? The industrial correspondent, on being invited by Sir Derek Ezra to attend a press conference, said that he 'was not doing PR for the Coal Board'. Margaret Thatcher, again at the press conference, inquired when the BBC make-up team would be arriving to prepare her. She was told very firmly that the BBC did not give make-up services for 'pressers' to anyone. Sometimes interviewees play hard to get and play off ITN against BBC. After the budget speech both Campbell Adamson of the CBI and Len Murray of the TUC were to be interviewed for the *Nine O'Clock News.* They had been present during the day at the BBC, involved in a current affairs programme, and were to be collected for their one-and-a-half minute insert interview for the *Nine O'Clock News.* Len Murray was immediately available but Adamson disappeared, causing much panic in the newsroom, and had to be rushed back from CBI headquarters for the interview.

People and opinions in the newsroom

News production is organised, as at ITN, by the division into intake and output departments. The terms are self-explanatory. Intake section is responsible for the collecting and co-ordination of incoming material – dominated by home news – and is under the day-to-day management of the news editor. Editor of the day is not a formal title but describes the function of whoever is responsible for each bulletin output. He is drawn from the ranks of six chief assistant editors. Foreign news has the parallel collecting and co-ordinating methods and its own editor.

The output department includes the newsroom and processing, journalistic and technical activities from which the news is shaped and packaged and is under the day-to-day control of the duty editor. The editor for television news and his two assistants (one of whom seems to have entirely administrative duties) have their offices down the hall and are seen in the newsroom from time to time, looking, in their shirt sleeves, very much like journalists. There are often announcements by one or other of them pinned to the notice board in the newsroom. I think it was my presence in the newsroom that may have prompted this sermon to the journalists who received it with something less than respect.

> To: All Regional News Editors. Copy to: A.H.R.T.D.,
> For information: Assistant Editors,
> Television News.
>
> Whichever way we look at it, I think we're committed to some very intensive coverage of industrial/political/social affairs for the foreseeable future. There are two ways of doing this – we can either sit back here in the Spur and do effective and elegant lantern lectures (complete with lovely slides and graphics); or we can report in a gutsy and straightforward fashion what these abstruse industrial/political decisions mean to the individual.
>
> If you reread the above, you'll gather which way I'd prefer we cover these events! Could you, therefore, bend your minds to considering how best we can report the effects of short-time working, of tax re-structuring, of oil development, of all the multitude of events we can expect in the way of Orders in Council and in Bills over the next twelve months or so.
>
> What I'm looking for is coverage of *people* – how are people faring at a time of economic anguish; abstractions are alright as pegs for stories, but everything in the end comes down to reporting events as they affect individuals.
>
> Think on these things, if you will, and when making offers to Alan Perry or the Assistant Editors, keep this kind of approach uppermost in your minds. 'People and Pictures' is the watchword for industrial coverage in 1975!

Administratively the newsroom is locked into the Corporation by the Editor of News and Current Affairs (ENCA). It is through his office that the full weight of BBC policy is implemented in the newsroom. People in the newsroom are very status-conscious and aware of their own and others' relationship to the complex grading system which covers employees of the BBC. Senior personnel, despite professional camaraderie, are defferentially treated by those beneath them. I am told that each grade in the upper hierarchy of the newsroom is given extra perks, i.e. television sets, hospitality privileges, and so on.

One of the senior editors clearly misses the conviviality and excitement of the newsroom. He comes in on quite slim pretexts to chat with people working here or to deliver to me homilies on the 'sound judgment' of newsmen, especially those employed by the BBC. His view is the classic apologia for television news. He considers that the news can only be a small selection of all possible events. However, 'the accumulative wisdom of seasoned journalists' was guarantee enough to ensure that television news carried only those items of interest and use to the largest section of the public. This editor was at pains to dismiss sociology as 'academic claptrap' or 'totally prejudiced rubbish', but had no problems at all in assuming that journalists were, by training, totally objective. I questioned the feasibility of objectively 'subjective' news, especially if one felt the audience to be of differing class composition. Somewhat surprised, he dismissed the idea of class altogether, saying 'we are all middle-class now'. I suggested that the news represented a false consensus view of the world to which he replied, 'it is a consensus of normality'.

Juniors (sub-editors) in the newsroom, commented on these views with a degree of cynicism. The 'accumulated wisdom of the journalist' became 'manufactured instincts' and some were of the opinion that the long-toothed journalist 'would not know the real world if he saw it'. One strange inversion of the 'man bites dog' principle of news value was from a senior deputy editor who reckoned that television news production was so routinised that it became easy to lose sense of the unusual. In fact, the unusual – events which did not immediately find their place in the news-making codes of production – could panic the newsrooms. One journalist said that the BBC had 'cocked up the assassination of President Kennedy – we couldn't decide whether it merited the death of a head of state procedure or whether it was just crime'. This is an interesting comment for two reasons. It justifies our choice of category system, again highlighting the way in which the newsroom insists on compartmentalising incoming information in terms of the rationality of news production. Second, it shows that the difficulties faced by the newsroom in the face of what are called *flashes* (urgent inserts into a bulletin) is not just a procedural difficulty but that uncoded inserts disturb the underlying sense of order that is part of news production.

Departmental meetings

Although the bulletins appear to be the product of the skills and co-ordinating machinery of the newsroom, it is only the visible iceberg tip of the total activity. Meetings, the daily and weekly futures meetings, both home and foreign, and most importantly the weekly NCA (News and Current Affairs) meetings, represent the crucial decision point in the structuralisation of the news.

I have been allowed into all except, significantly, the weekly NCA meeting. It is here that the overall scrutiny and policy of the Corporation is expressed. It is attended by upper management of the BBC and by only senior representatives of the newsroom. From here issue the logs and memoranda that make up the 'manufactured instinct' of BBC journalism. Directives as to style and nomenclature are issued from here. Semantic usage is clarified. Decisions, for instance, as to whom is to be designated terrorist, guerrilla, or even freedom fighter are made at this meeting. More important policies such as that governing the coverage of events in Northern Ireland or Portugal are received from the top echelon of BBC management and are issued down to the newsroom.[12]

The decisions taken at this meeting are tangible in the day-to-day handling of news. Often during the course of the daily meetings I attended time would be spent applying the precise definition of the NCA directive on 'fair and impartial coverage of the referendum campaign'. It seemed to be a problem whether impartiality and fairness was to be interpreted as equal time allocation to the pro and anti groups or 'on the one hand, on the other' impartiality within items. Most of the journalists seem to fret at this imposition on their journalistic skill. Always the final decision is the responsibility of the editor of the day.

Finally I went to the weekly futures meeting for Saturday 19 April. It was attended by the Editor of Television News and all duty editors who would be involved in the day's coverage, as well as correspondents, technical managers, and sub-editors in charge of special features. Predictable or fixed events were then offered up to the correspondents or duty editors for consideration and the possible modes of coverage discussed. By and large there was a high degree of journalistic agreement. It seemed clear that the Editor of Television News controlled the meeting, in so far as from his vantage point as chairman he signalled quite clearly the importance or otherwise that he attached to events and their possible manner of coverage.

Such meetings as these are crucial, not only because of the continuing formulae from which news is produced but also as a legitimation among professionals of the criteria with which their judgment is bound up. Decisions seem to be based upon the following criteria (although it would be necessary to sit in on more than one of these meetings to even begin to see what is being negotiated and in what manner). The most immediate criterion is that of shared journalistic notions of news, the supposed result of accumulated experiences in similar news-making enterprises. A demonstration by authors over the library borrowing and payment scheme was accepted for coverage because of the potentially 'unusual' nature of the demonstrators. Another criterion is the identification of this particular news operation with the public image of the BBC – 'we do not like our reporters to door-step', i.e. waiting around for unscheduled interviews. The Queen and members of the Royal Family at a chapel service conducted by the Archbishop of Canterbury was accorded an outside broadcast unit coverage – perhaps a residue of the BBC's role as supplementary court circular.

A third criterion of importance is that events should be selected and interpreted in line with overall BBC political policy. 'Hitler's birthday being celebrated in Sussex' was obviously a possible item; however, the policy concerning extreme political groups precluded this event immediately. Kenneth Littlejohn's appeal against extradition order, Division Court, London, was greeted by silence and a firm 'I think not' from the Editor of Television News. Further, there is a self-conscious interpretation of the BBC's responsibility to inform the public which is sometimes in contradiction to the middle-class attitudes these journalists appear to have. On the one hand a lurid murder trial was greeted with 'we're not a bloody tabloid', while a 'Black Paper on Education, already widely leaked' was accepted since it had been leaked in 'posh papers'.

Finally the shared middle-class perspectives of the journalists become very clear in their attitudes to such things as strikes – 'just another one to record', 'more of the same blackmail', and so on.

The whole meeting had the atmosphere of head boys and prefects or regimental pow-wow (re-emphasised by the fact it is an all-male assembly), which was most clearly expressed in the kind of humour employed.

One item, a festival of art and music in West Yorkshire to be attended by the Prime Minister, was dismissed as a 'fish and chip' festival (laughter and applause). The launch of the Fiat economy car was greeted with 'made with slave labour in Yugoslavia' (cheerful assent). The conference of National Federation of Business and

Professional Women's Clubs caused quite a lot of sexist comment – 'we can't get that many trendy bottoms in' and so on.

Given that this meeting provides an ideal opportunity for re-assessment of performance practice, examination of mistakes, complaints and current policy, it is surprising that there was no indication of such activity. Indeed the full sense of energy of the news team functions to inhibit argument, or collective criticism – and this meeting proved no exception.

Social control

Without doubt, the weight of the BBC's tradition and organisational authority weigh very heavily on employees. The pressure to conform to the BBC ethos is constantly maintained in many ways. All kinds of benefits and privileges accompany the upward path through the hierarchy. Breaches of conformity can be punished by removal to obscure managerial shuntings in the case of those thought to be dependable, or for the creative 'mavericks' a removal to Current Affairs or some more expansive department. Over all, at this time, hangs the real threat of redundancy. I spoke to one or two who thought that potential nonconformists were noted by the management – 'they turn down the left-hand corner of your file' – 'and you get discreetly pushed'.

Here, as at ITN, it was clear that apart from the few who had day-to-day control and power, journalism in a television newsroom can be very boring and unfulfilling. The processes, the stories themselves, become endlessly routine, the easiest excitement again being that of beating the clock and getting yet another bulletin out on time, with as few hitches as possible. Sometimes excitement breaks out, as when a reporter danced across the opening shot of the *Nine O'Clock News* in the newsroom. He was severely reprimanded and all were generally warned against 'horseplay'.

4 MEASURE FOR MEASURE

> If Moses came down from Mount Sinai with the Ten Commandments in an era of television, he would certainly be greeted by camera crews.
>
> 'What do you have?' they would ask.
>
> 'I have the Ten Commandments,' replies Moses.
>
> 'Tell us about them but keep it to a minute and a half,' they would say. Moses complies and that night on the news, in still more abbreviated form the story is told. The newscaster begins, 'Today at Mount Sinai, Moses came down with the Ten Commandments, the most important three of which are. . . .'
>
> William Small

A number of observers have noted the 'sameness' of the news from day to day in terms of structure, content and presentation. Newspaper reporting has been criticised by Rock for its ritualised and cyclical nature which conveys an impression of occurrences – and society – as 'eternal recurrence' with no grand design.[1] Frayn has parodied the formal bureaucratic nature of news production as well as the less formal interpretative rules, imagining a computerised system of news production which would break 'its last residual connection with the raw, messy, offendable real world'.[2]

These mosaic patterns of the newspaper have their parallels on the screen. The following analyses can thus be regarded as using similar computer techniques in reverse, a process of unravelling the news bulletins to reveal some of the codes of story classification, placement, duration and presentation which are routinely employed. The results indicate that, given a range of stories from which to choose, they will be contained within a clearly defined structure and will receive predictable treatment. The codes of news production create media packages so obvious as to be instantly available for parody but this does not necessarily mean they are understandable. In this context, familiarity should provoke questions rather than complacency. The last chapter indicated that television news producers are relatively unaware of the ways in which they allow their routines of news processing, the effects of time pressures and their need to compete for audiences, to force their output into predictable patterns.

Bulletin profiles

Length

Broadcast news has travelled a long way since the BBC transmitted a single radio news bulletin composed by the wire service and only read after 7.0 p.m. (because of the hostility of newspaper proprietors). Today the regularly scheduled news bulletins vary in length from a few minutes to half an hour – or more in certain circumstances. On weekdays both BBC1 and ITN carry bulletins at lunchtime, early evening and at 9.0 and 10.0 p.m. respectively. The *Nine O'Clock News* and *News at Ten*, which are regarded by their producers as being competitors, are the main programmes on these two channels, with a well established format. ITN's *News at Ten*, which began in 1967, is usually 25 minutes in length (excluding titles, openings and closings, which take up to one and a half minutes). The *Nine O'Clock News* is slightly shorter, lasting about 20 minutes. In the early evening, the BBC1 programme (12 minutes) is slightly longer than ITN's (10 minutes). *First Report*, ITN's lunchtime programme, lasts 20 minutes, during which the presenter regularly conducts his own live interviews in addition to the more conventional coverage.

The lunchtime bulletin on BBC1 lasts for less than 3 minutes and has much in common with the summary included in BBC2's *Newsday* programme and the shorter weekend bulletins. *News Extra*, the late news bulletin on BBC2, at the time of the study was more variable in length than other bulletins but normally lasted about 20 minutes. The exception to the generally shorter weekend bulletins is the *News Review* programme, for the deaf, on BBC2 lasting for 40 minutes. This differs from the other programmes in taking stories for the whole week and presenting them with subtitles for the benefit of the deaf. It is transmitted early on Sunday evening.

The length and scheduled time of the news is a product of the historical place of the news in television output as a whole, the daily exigencies of scheduling, and the need to cover extraordinary events. Of these, the first is the most important. Legal restrictions on the hours of broadcast news have long since ended but restriction of the hours of broadcasting still exists. Economic pressures are at least as important. But whatever the hours of broadcasting, the proportion of news and news-magazine programmes within them has remained very stable. The increase in the length of the main evening news programmes has to be seen against the background of an increase in total output. Since 1956, commercial television has included between 6 and 10 per cent of news and news magazines, although the total ITV output has more than doubled in the same period. The BBC carries a similar amount.

It has been argued that the proportion of news in the schedules has less to do with institutional structures of ownership, legal controls or even the policies of broadcasting organisations than with economic pressures and the quest for audiences.[3] However the proportion is to be explained (and the public nature of the BBC confounds a simple economic rationale), a formula has been settled on which seems to have its own justification in terms of scheduling. Any extension of the news is seen by the institution as upsetting this formula and therefore only appropriate to minority times on a minority channel.[4] There are sometimes exceptions to this arising from other scheduling requirements which occasionally encroach on news time. The third set of circumstances – the need to cover extraordinary events – occasionally results in an extended news. This occurs more frequently on ITN than BBC, despite the greater difficulties of liaison with fifteen commercial companies, each with its own programming. Although short extensions are not infrequent, there is actually a greater chance of the bulletin being shorter than the regularly scheduled length. The extended news is a rare occurrence. ITN's coverage of the outcome of the Conservative Party leadership campaign was the longest extension in the first three months of 1975, a three-part programme running for 45 minutes.

The number and duration of items

The number of items in a bulletin does not have a normal distribution, statistically speaking, from day to day. Except in the shortest bulletins, one cannot predict any single number of items as being more likely to occur than another. There are, however, certain regularities between channels and between bulletins. Consider the mean number of items in each weekday news programme and the range and type of variation in this mean (Table 4.1).

Table 4.1 Number of items in each bulletin (weekdays)

	Mean no. of items	Modal value[1]	Range[2]	Modal length[3] mins secs
BBC1 Lunchtime	5·6	6	5	2.30
First Report	7·1	8	9	21.00
BBC1 Early Bulletin	10·5	9	13	11.30
ITN Early Bulletin	9·3	10	11	10.30
Newsday	6·8	7	9	2.30
Nine O'Clock News	12·2	11	15	23.30
News at Ten	14·1	12	16	25.00
News Extra	5·9	4	9	18.30

[1] Most frequently occurring number
[2] Difference between smallest and largest number of items
[3] Most frequently occurring length of bulletins

In general, the mean number of items, the modal number of items, and the range increases steadily with the length of the bulletins. The longer the news the more stories are likely to be included, because longer bulletins do not include, as would be possible, fewer longer items. A total of less than four or five items seems to be unacceptable for bulletins of more than 3 minutes in length. However, the bulletins are subject to changes in presentation. In the period of our study, two programmes were moving away from this norm. *First Report* and *News Extra* tended to have fewer items, a smaller range of items and more editions with a very small number of items. Both have negatively skewed distributions of the number of items in contrast to the positively skewed distributions of the other bulletins. This skewing is the statistical expression of the fact that these two bulletins were becoming more like current affairs programmes.

In cases of bulletins which are of comparable length and which are scheduled at similar times (the two early bulletins, for example, and the *Nine O'Clock News* and *News at Ten*), there is a close similarity between channels. The slightly greater range and average number of items in the BBC Early Bulletin and *News at Ten* respectively, is connected with their slightly greater average length.[5] Although the BBC1 lunchtime and BBC2 *Newsday* bulletins are normally the same in length, the higher values for the *Newsday* bulletin can be accounted for by a handful of bulletins which were extended beyond this time and had considerably more items. These extensions are because the format of *Newsday* (news bulletin and current affairs interview) facilitates such extension without altering the overall running time. These and other comparisons show that bulletin formats do not vary in random ways but have a similar underlying logic.

This provides some evidence against the argument for an extension of news time as a means of correcting the 'bias against understanding'. In the present situation, all that can be said is that the longer the news, the greater the number of stories covered.[6] It cannot be said that stories are treated at significantly greater length or in a significantly different way. Instead of the longer news bringing a greater element of explanation and commentary to roughly the same number of items – and therefore stories – the effect of lengthening the news is to extend the number of stories. There is one caveat to this conclusion. Despite a greater number of items in longer bulletins, the average duration of items varies from half a minute in the *Newsday* summary to 3 minutes in the longer programme on BBC2 (Table 4.2). Thus, there is a general increase in average item durations

Table 4.2 Mean duration of items in each bulletin

	Mean item duration Weekdays mins secs	Weekends mins secs
BBC1 Lunchtime	0.35	0.50
First Report	2.40	—
ITN lunchtime	—	2.15
BBC1 Early Bulletin	1.15	0.55
ITN Early Bulletin	1.00	1.00
Newsday	0.30	—
BBC2 *News Review*	—	2.55
Nine O'Clock News	1.40	1.30
News at Ten	1.35	1.30
News Extra	2.50	3.15

with the increase in bulletin lengths. But this increase still leaves the basic number of items deemed necessary for a longer bulletin unaffected. There continues to be a close similarity between the two early bulletins on the one hand and the *Nine O'Clock News* and *News at Ten* on the other. *First Report* and *News Extra* are again exceptions to this, for the reasons outlined above.

How the increasing duration of items is to be reconciled with our finding of a fairly constant number of items in bulletins of different lengths, is shown in Figure 4.1. Essentially the length of items in a bulletin does not keep step with the number of items in the bulletin,

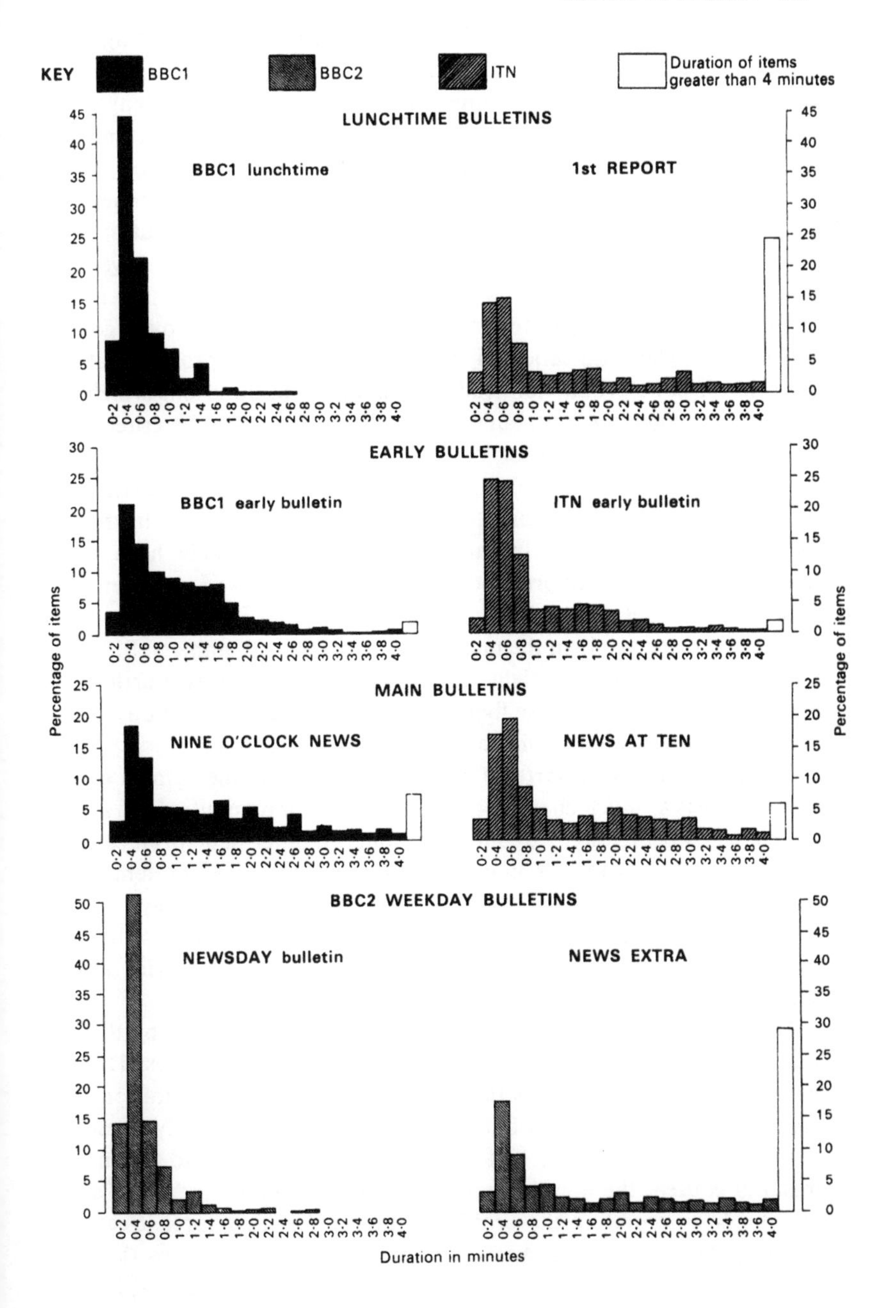

Figure 4.1 Item duration profiles (weekday bulletins)

although the mean length does increase. Each bulletin has a single peak in Figure 4.1, i.e. a high proportion of the items transmitted during that bulletin will be of one particular length; but this length is different for each bulletin. With *First Report* and *News Extra* this is still true although the peak is lower, the larger proportion of longer items reflecting the current affairs style of these programmes. The profiles in Figure 4.1 show that in each case the distributions have a single peak in the range of about half a minute. This is even true of the longer bulletins which thus use the extended time to include a greater range of stories as much as more detailed coverage.

There is a clear indication in the similarity of these profiles that the rules of news production lead to a mix of item durations which does not vary greatly from bulletin to bulletin. Whether the criteria are journalistic (news value or 'covering the waterfront') or aesthetic (the pace of the programme) the mix will include a high proportion of items of less than one minute. Half or more of the items in the majority of bulletins are in this category, although ITN have a significantly higher proportion than the BBC bulletins (68 per cent and 56 per cent respectively in the early bulletins and 52 per cent and 46 per cent respectively in the main evening news programmes). BBC bulletins therefore have a higher precentage of longer items. Further analysis will show that these figures conceal significant variations in the average duration of items in different story categories.

In the case of *First Report* and *News Extra*, whatever the number of items, there is as much chance of an item being 4 or 6 minutes long as there is of its being, say, 1 to 3 minutes long. In other bulletins it can be assumed that, with certain exceptions, the duration of an item is the best indicator of its attributed importance. This is probably true as well of *First Report* and *News Extra* with the qualifications that both programmes seem to aim to cover a few salient stories at greater length and usually with interviews.

Originally we felt that item durations would be governed by the use of film, graphics or interviews, giving steps in these profiles. The lack of the sharp breaks, therefore, is evidence that item durations are entirely flexible. A lack of film material does not necessarily mean that an item lasts less than half a minute. On the other hand, if film material is used, the item is almost certain to be longer than half a minute. In interpreting the profiles in Figure 4.1, therefore, it can generally be assumed that apart from newscasters, short items (less than 1 minute in length) will contain pictures, maps and other graphic inputs but rarely film; that items between 1 and 4 minutes will usually contain film as well as graphic inputs but not interviews; that longer items will contain a whole range of technical and personnel inputs.[7]

Preliminary cross-national studies of television news indicate that the profiles shown here are not untypical of the majority of European bulletins. Rositi, for example, has demonstrated that news items in Italy, Germany and the UK, last little more than a minute and a half on average and that two-thirds or more of the items are below average in length.[8]

French television was the only significant departure from this norm in Rositi's four-nation study. He also encountered greater difficulty at the coding stage in deciding what constituted an item on British television news. A greater proportion of stories were run together by scripting – a clue to the greater propensity to create 'package' items. This does not occur with sufficient frequency, however, to replace the prevailing logic of news as fragments of information with little apparent overall coherence apart from that imposed by the bulletin format, which utilises and expresses the interpretive frameworks of the producers. The average duration of items and the amount of packaging varies according to story category and channel. The variations indicate certain priorities in news gathering and news processing. They also reflect certain organisational arrangements in the newsrooms. Table 4.3 shows the average

Table 4.3 Average duration of items by story category and channel (all bulletins)

Story category[1]	Average duration of items BBC1 mins secs	 BBC2 mins secs	 ITN mins secs
10 Politics	1.35	2.00	2.00
20 Industrial	1.10	1.25	1.40
30 Foreign	1.15	2.00	1.35
40 Economics	1.40	1.00	1.15
50 Crime	1.10	0.55	1.30
60 Home Affairs	1.25	2.25	1.35
70 Sport	0.50	1.55	1.10
80 Human Interest	0.55	1.55	1.00
90 Disasters	1.05	1.15	1.10
00 Science	1.15	3.00	1.20
Av. duration (all categories)	1.15	1.45	1.25

[1] A full description of the classification system for these categories, as well as the sub-categories not shown here, is given in Chapter 2, and in Appendix 1.

duration of items in each of the main story categories. The greater overall average in ITN and BBC2 is partly accounted for by extended programmes such as *First Report*, *News Extra* and *News*

Review, whose distinctive profiles we have noted. In the case of ITN a greater tendency to run items together in certain categories may be another factor which contributes to the higher figure.

A common-sense assumption that some areas, such as economics, require greater time because of their complexity, is not borne out. Even if the relatively short stock and currency market reports are taken into account, at most (on BBC1) 50 extra seconds on average are devoted to economics as against, say, sport. On the other channels it is even less and in any case, the duration differences we are here dealing with are quite small. This view must be tempered by the common professional assumption that the compression of information on television means that 50 seconds is a long time. The durations nevertheless reveal that no real presentational differences are possible, since there is a maximum of 50 seconds to play with, as it were.[9] It is also clear that the expected skew in average durations, with longer items being more common in 'serious' categories, does exist – politics for instance, are the longest items on average. All three channels show a similarity at the other end of the scale. Items in the categories of crime, sport, human interest and disasters, all tend to be shorter than average, a feature which is associated with their low priority in terms of placement.

There is one significant difference between channels. The industrial and economic categories do not have the same average item durations across channels. The figure for ITN in the industrial category is significantly above the comparable figure on both BBC channels (Figure 4.2). The picture is reversed in the economic category. BBC1 has the highest average item duration.

ITN carries a significantly higher proportion of lengthier industrial items and the actual number of items is smaller. This is counter to the general tendency for BBC items to be longer. The percentage of industrial items longer than two minutes varies from 18 per cent on BBC1 through 25 per cent on BBC2 to 32 per cent on ITN. This evidence of a greater amount of 'packaging' by ITN leads us to initially suspect differences between channels in the process of news gathering and processing itself rather than a simple running together of stories for elegance of style or smoothness of presentation. The organisation of industrial news gathering includes one major difference between the BBC, who had one specialist industrial, and one economic, correspondent and ITN which had a team of three specialists in industrial, economic and business affairs. This organisational difference directly accounts, we would submit, for ITN's

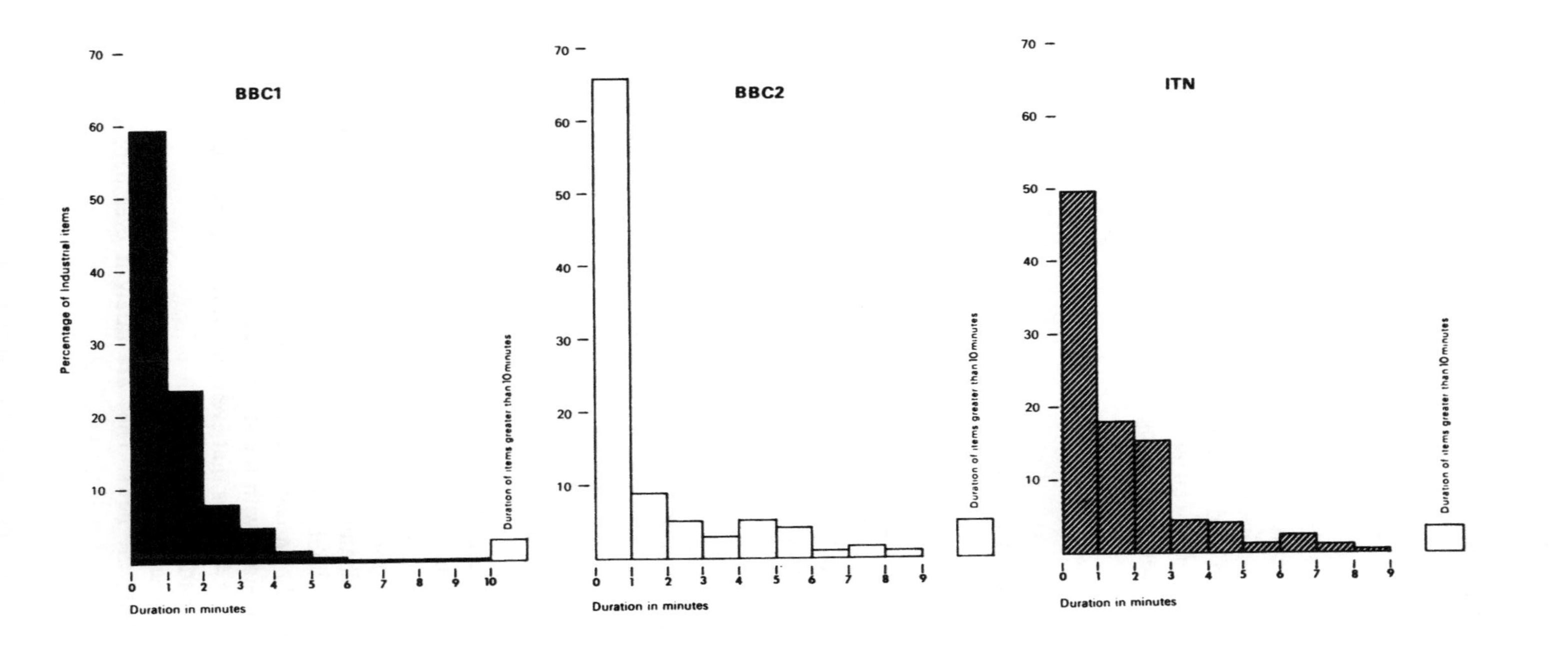

Figure 4.2 Comparison of industrial item durations

greater tendency to package items in the industrial area, since there is a higher input of specialist reporting. The industrial correspondent appeared to camera in only 30 (7·2 per cent) of BBC1's industrial items, whereas the three ITN specialists appeared in 49 (12·2 per cent) of industrial items on that channel. We are not however suggesting that the greater use of specialists necessarily leads to a parallel increase in cohesion and comprehension. The package is only a presentational device. It can as easily create spurious causal connections as it can lead to an integrated view of the world.

Category profiles

Distribution of items by category

We analysed the content of the news according to categories as described in Appendix 1. The number of items in each of these categories is similar from bulletin to bulletin and channel to channel. The relative ranking as shown in Figure 4.3, remains basically the same for each channel. All three profiles, which indicate the percentage of items in each category, have a distribution which is more like that of the 'quality' than the 'popular' press. A large majority of the items are devoted to political affairs at home and abroad (interpreted as a rule through the activities of political leaders – hence the emphasis in our period on the Conservative Party leadership issue); industrial topics and matters of social and economic management. We shall describe these categories as being at the 'serious' end of the category spectrum. The proportion of time devoted to sport, human interest and other non-public affairs coverage is in the order of 20 per cent – less than even the quality press, in which the corresponding figure is over 30 per cent.[10] This is the opposite end of the spectrum. The relative predominance of the foreign, political and industrial categories is the same for all three channels, industrial items taking third place. Even in the less prominent areas of coverage, the order varies little from channel to channel. For example, there is a similar amount of slightly more economic news (41–44) than home affairs news (61–65) on each channel and (apart from BBC2) this similarity is maintained with the sport, human interest and disaster categories. Figure 4.4 shows that even within these broad categories the allocation of items is identical.

The priorities revealed here are obviously deeply embedded in the practice of making the news. They are rarely if ever debated or

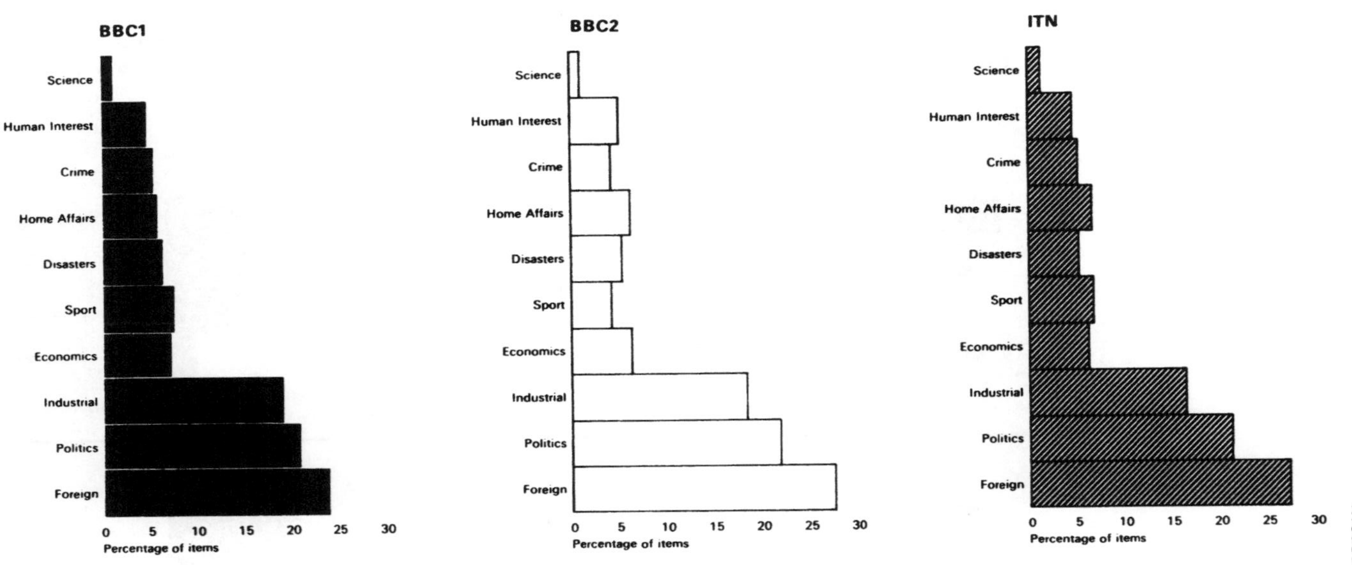

Figure 4.3 Content profiles for each channel (includes all bulletins) showing the percentage of items in each category

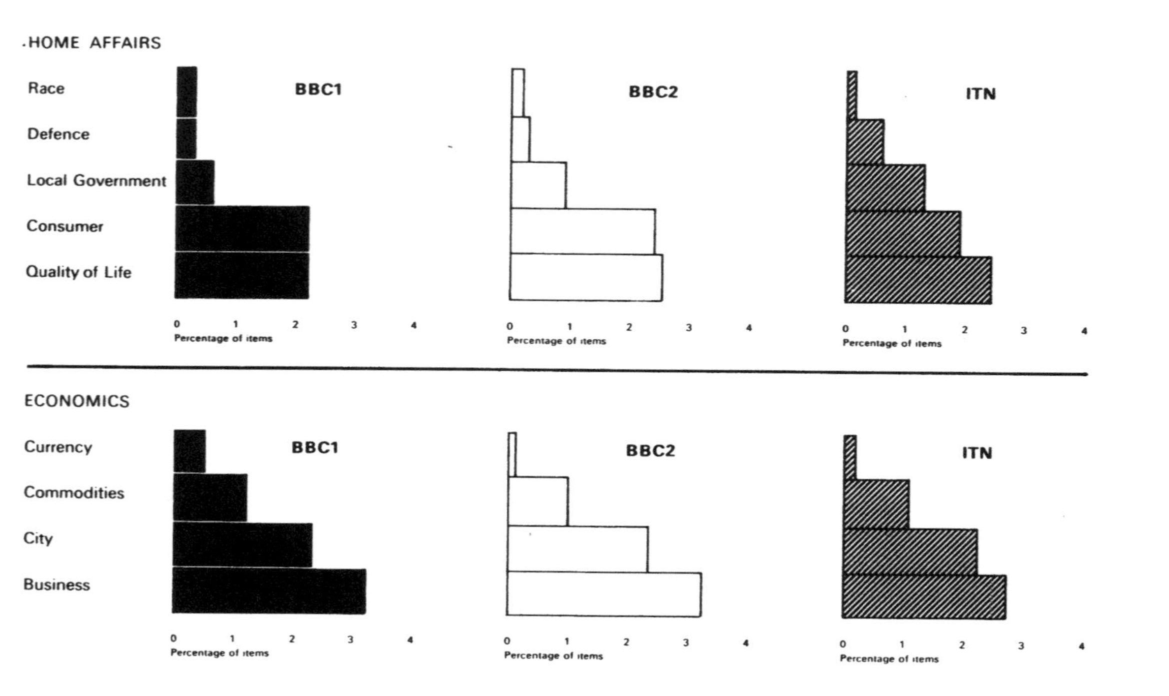

Figure 4.4 Economic and home affairs content profiles for each channel (includes all bulletins), showing the percentage of items in each category

justified, whether in terms of audiences or patterns of activity in the world. Yet why, for example, should items on wages, disputes and employment in industry be presented with far greater regularity than items on investment, profits and other aspects of industrial management? Or what is the rationale for the proportion of human interest and 'feature' stories? If, as is most likely, this mix has been inherited from the press – and a minority section of the press – it is interesting to note that the formula is nevertheless used to create a major source of news for the majority of the population.

For most categories, variation between channels is in the order of 1 per cent or 2 per cent of the total number of items. In the larger category of political news (11–19) this variation is still very low. Grouping all the sub-categories, the percentage of political items is 20·6 per cent, 21·8 per cent and 21·1 per cent on BBC1, BBC2 and ITN, respectively. The only significant exceptions to this general similarity of profile (see Figure 4.3) are, first, the greater predominance of foreign items on both BBC2 and ITN, and second, the relatively low percentage of sports items on BBC2. The explanation for this lies partly in the programme philosophy of *News Extra* discussed above, early transmission time of *First Report*, and partly in variations of bulletin length between channels.

We did not anticipate such a high degree of similarity. It would not be surprising to find this similarity if the news were manufactured in identical institutions with common resources and aims. But the institutions have slightly different structures and markedly different atmospheres. Their processes of news gathering are independent and, not least, they are assumed to be in competition and take pride in extremely subtle marks of distinction.

Taking each bulletin separately, such differences as do exist in the mix of categories have chiefly to do with the variation in bulletin length. Thus, instead of a short bulletin having a category profile which corresponds exactly to the average profile for that channel, it tends to have items from a smaller range of categories and more items in the 'serious' categories. The differences remain quite small, however, as Table 4.4 shows.

Catering for different audiences at different times of the day (e.g. women's programmes in the afternoons) is a norm of television scheduling. Table 4.4 illustrates that there appears to be no orientation towards the audiences, which is surprising in view of the fact that much of the internal research of both BBC and IBA has shown significant differences in audience profiles. These are differences of

Table 4.4 Distribution of items by bulletin (weekdays), showing percentage of items in each category

	BBC1			BBC2		ITN		
	Lunch-time	Early	*Nine O'Clock News*	*Newsday*	*News Extra*	*First Report*	Early	*News at Ten*
10 Political	21·1	22·0	21·6	23·6	25·4	23·6	22·3	21·1
20 Industrial	17·9	16·6	16·6	18·6	15·3	14·3	16·6	13·1
30 Foreign	24·7	21·9	21·1	20·7	25·7	27·3	24·3	25·9
40 Economic	5·2	8·3	9·8	9·7	6·9	7·1	7·7	7·6
50 Crime	4·3	4·8	4·5	5·5	2·8	4·3	5·2	5·1
60 Home Affairs	5·2	6·6	6·9	5·8	6·9	5·6	5·4	7·2
70 Sport	7·1	7·6	7·5	5·0	4·1	6·5	6·0	7·4
80 Human Interest	4·6	5·3	5·4	4·2	5·1	3·2	4·7	6·6
90 Disasters	9·9	6·1	5·1	6·3	5·8	6·7	5·9	4·4
00 Science	—	0·7	1·5	0·5	2·0	1·3	1·8	1·6
Total %	100·0	99·9	100·0	99·9	100·0	99·9	99·9	100·0
n =	324	685	826	381	393	462	613	942

size, social class, age and gender as was shown in Chapter 1. Only one bulletin is specifically targeted – *News Review*, designed for the deaf.

When examining the exceptional bulletin formats of *First Report* and *News Extra* we find the profiles of categories, although different from the other bulletins, are not seemingly targeted to housewives or, say, tired intellectuals. *First Report* differs from other ITN bulletins in having a higher proportion of foreign items (in fact the highest proportion in any bulletin, both ITN and BBC) and a higher proportion of Northern Ireland news. These is also a greater proportion of disaster items. The proportionally smaller categories are human interest, politicians (personal) and science. The lunchtime audience has a high proportion of those housebound through age or family roles but the priorities of *First Report* do not appear to take this into account in any special way. Although Robert Kee's style of presentation may be an adaptation to the lunchtime audience the high proportion of foreign news, disaster stories and Northern Ireland stories (which frequently deal with terrorist incidents), can be accounted for without reference to the audience. *First Report* is the first major television news programme of the day and as such is likely to concentrate on the major stories which have broken early enough to be included. Many of the important but routine activities of politics, business and industry take place throughout the day and so cannot be fully reported at lunchtime. Untimely events such as disasters or news from different time zones abroad or news which is obtained from the EBU's first news exchange, therefore stands a greater chance of inclusion.[11]

News Extra caters for an audience of a different kind. Does its composition affect the range of story categories used? There is certainly no fundamental re-ordering of these priorities; the overall profile is similar to all other bulletins. As might be expected of a minority channel bulletin with longer than average items, there is a tendency to have somewhat fewer crime, sport and human interest stories and more in the political, foreign and science areas. What is significant for us is that there is no corresponding increase in the industrial news category. With only 15·3 per cent of industrial items this bulletin has one of the lowest proportions. It would appear that an opportunity for extended coverage of these stories was not taken up in *News Extra*, which concentrated more on political[12] and foreign news.

There is therefore little to suggest that the audience affects the priorities of television newsmen. These are broadly similar regardless of age, sex or class differences among the audience. The mix of news for any of these audience groups is the same as the news for any other group in a way which is quite uncharacteristic of other news media. The variations which do exist can be explained by reference to the time of day and to other factors which we will now consider.

Relative duration of categories

We have already noted that there are significant channel differences in the average length of industrial items. This is true also of categories other than industrial, although the differences are not so great. The distribution of items by category therefore has a profile which differs in certain respects from the profile which represents the allocation of time. Figure 4.5 provides the means for a systematic comparison of these duration profiles with the distribution of items which was shown in Figure 4.3.

In certain categories on BBC1 and ITN, the percentage duration is lower than the percentage of items. These are the categories of economic (City), crime, sport, human interest, disasters and science. The majority of political stories on the other hand, have a significantly higher percentage duration, especially in category 16 (Conservative Party), which attracted a large amount of 'feature' coverage during the leadership election. On ITN, category 19

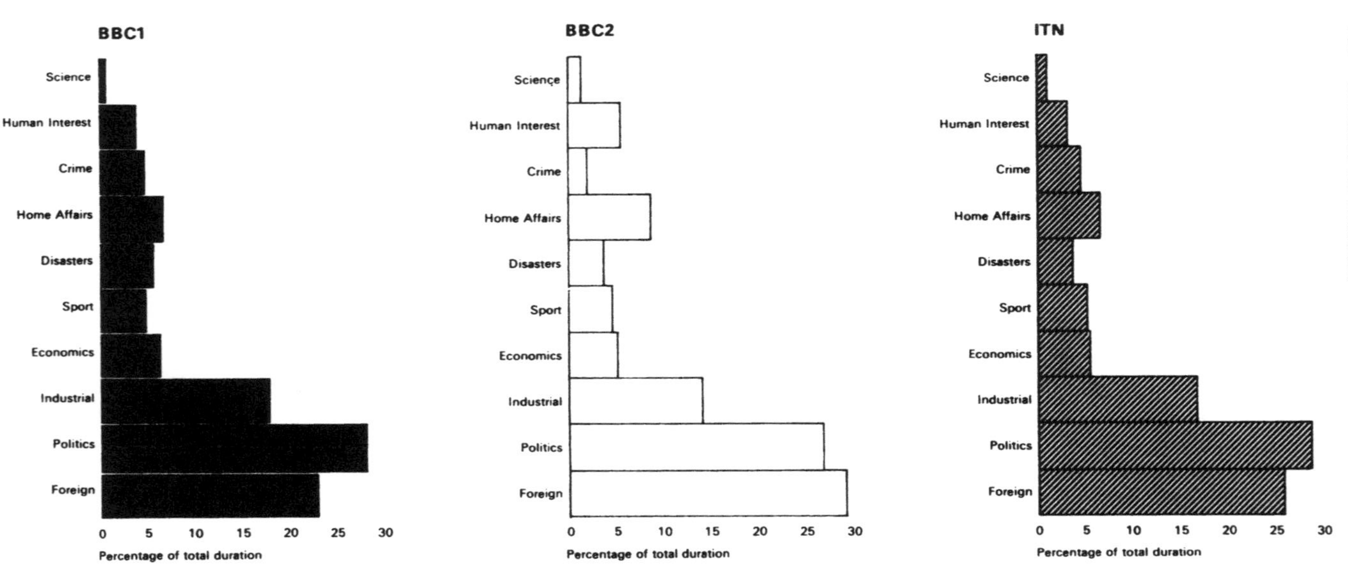

Figure 4.5 Content profiles for each channel (includes all bulletins) showing the size of each category as a percentage of total duration

(politicians, personal) had a smaller percentage duration because these items have much the same function as human interest stories. In the same category on BBC1 the relationship between the number of items and duration is reversed because BBC made much of their exclusive Australian coverage of the Stonehouse story, often running it as a lead. Political news in general also has a higher percentage duration on BBC2 but, in contrast with the other two channels, sport, human interest, disaster and science stories actually received more time than would be anticipated from the count of items.

The most striking between-channel difference is in the relative number of items and the time devoted to them on BBC1 and ITN in the industrial and foreign categories. The differences are displayed in Table 4.5 expressed as percentages of total bulletin items and

Table 4.5 Comparison of industrial and foreign coverage (all bulletins)

		Percentage of total no. of items		Percentage of total duration
Industrial	BBC1	18·5	>	17·4
	ITN	16·1	<	16·6
Foreign	BBC1	15·7	>	15·0
	ITN	19·2	>	16·7

durations. In each case the number of items allocated to the category is greater than the percentage of time devoted to them, except that ITN industrial items in contrast last longer than a simple item count would suggest. Thus, although individual industrial items are longer on ITN, the gap between the industrial coverage in general on BBC1 and ITN, which is apparent in the item profile, is narrowed considerably in the duration profile. The same is true with the foreign category.

This illustrates the limitations of any content analysis which looks at only one or a limited number of measures of 'content'. For instance, ITN's use of a large proportion of foreign items does reflect a basic journalistic concern to 'cover the waterfront' and aesthetic criteria in providing a contrasting and varied mix of items.

Story categories: sources of variation

The flexibility of categories

Standard deviations of the percentage size of each content category can show significant differences of 'flexibility'. They show the chances

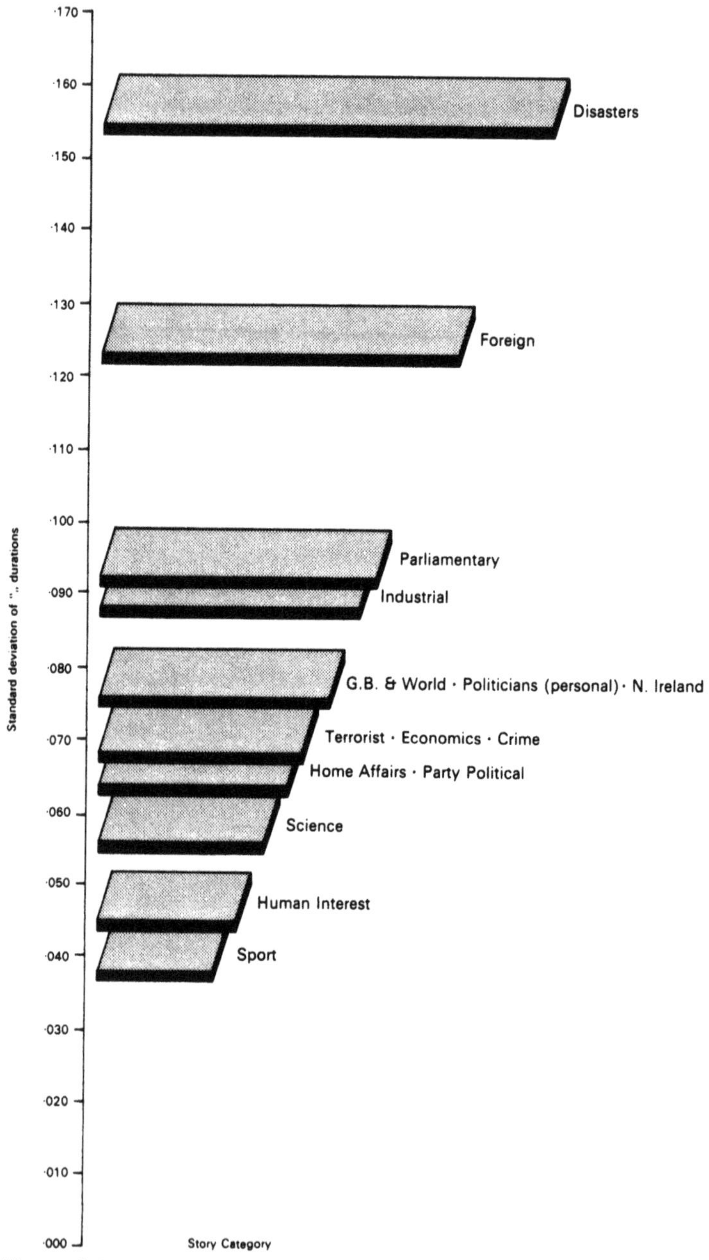

Figure 4.6 Story categories: dispersion of percentage durations (all weekday bulletins)

of any category occupying a much larger or a much smaller proportion of the news than the average for that category.[13] The ranking in Figure 4.6 indicates that there are categories with high, medium and low flexibility. The relatively low dispersions are sport and human interest stories. The amount of space they occupy in the news varies within narrow limits. This probably reflects the extremely routinised reporting of sports except on Saturday – which reduces them to something like a results service. They have low priority in the running order in line with the general 'quality' mix of the bulletins.

Disaster and foreign stories have far higher dispersions than other categories. As we will show this as a result of at least two story types in the disaster category. Major accidents and disasters within the UK are reported at length; most get rather brief treatment. (As journalistic legend has it 'small earthquake in Peru, not many dead'.) Evidence of item placings in fact shows that there is a strong tendency to put disaster stories either at the beginning or near the end of a bulletin; but not in the middle.

Foreign items have the next highest standard deviation. This again indicates that they have a relatively broad range. Theoretically the range of foreign stories is very large indeed and by no means are all of major importance. Other criteria apart from the importance of the story itself can be used to account for this. The availability of good film material will be a prime consideration.[14] Otherwise, some foreign stories are of world consequence (summit talks, Arab–Israeli diplomacy, the fall of Saigon) and are used as leads. Others are major running stories (civil strife in Beirut, Portugal and the future of Angola), which are most likely to be found in the body of the bulletins. Yet others are minor foreign items (art thefts in Italy, Fijian miners' sex breaks and riots in Peru) which are used like human interest stories – that is, to pay off a bulletin. Compared with industrial coverage, for example, foreign stories included both a higher proportion of short items (less than 1 minute in length) and a higher proportion of long items (longer than 4 minutes).

With the exception of parliamentary and industrial news, the remaining categories are clustered within a very narrow range. There is a tendency for most of these to be stories which are only infrequently dealt with at length or 'packaged'. These do not usually have a high priority in story placings. Figure 4.3, of course, only provides a measure of flexibility. It says nothing about the relative size or priority of the story categories. Crime stories, for example, occupy a small amount of time and tend to have low priority but they are used with about the same flexibility as Northern Ireland stories, which often have high priority and which form one of the largest political categories.

The slightly higher flexibility of parliamentary and industrial stories suggests that they are frequently packaged, but equally separate stories can appear in different places within a bulletin. Thus a car industry package will typically appear with other, separate items on, say, the textile industry, rail workers or teachers, but a Northern Ireland package is more likely to include all the stories from the province for the day.

We have shown that over a three-month period there were significant variations in the usage of different story categories. This confirms that certain categories occupy a very stable position within the news bulletin and that other categories may be used at a variety of different lengths and at different points in the bulletins.

Time of day

There are smaller scale variations in the size of some content categories which arise from the process of news gathering itself rather than from the conventional weighting of categories or general priorities. These small variations occur on a daily basis between early and late bulletins. For instance, there is a clear tendency for disasters to have relatively more coverage earlier in the day (a finding which cannot be explained simply by the difference in bulletin length). With the exception of *News Extra* there is a steady decline in the amount of disaster coverage from 10 per cent of all items on the BBC1 lunchtime bulletins to 4 per cent on *News at Ten* (see Table 4.6). The explanation of this lies as we have said, in the

Table 4.6 Variation in selected story categories according to time of day (weekday bulletins, percentage of items)

	Disasters	Parliamentary	n
BBC1 Lunchtime	9·9	2·5	324
First Report	6·7	4·8	462
BBC1 Early Bulletin	6·1	4·2	685
ITN Early Bulletin	5·9	3·9	613
Newsday	6·3	5·0	381
Nine O'Clock News	5·1	4·7	826
News at Ten	4·4	3·9	942
News Extra	5·9	6·6	393

fact that many stories in the main categories of politics and industry, based as they are on news conferences or the events of the day, have not occurred and therefore cannot be fully reported until the evening. Disasters, on the other hand – the majority of them at least – are by definition random.

Political events have their own logic. As would be expected, parliamentary coverage shows some tendency, especially on BBC, to become fuller as the day proceeds. *First Report* is slightly exceptional here with a higher proportion of parliamentary items than would be expected because there is frequent use of commentary by experts or participants on the previous day's business and forthcoming debates. There are some indications (see below p. 220) that early coverage of this kind is sought after by some public figures, including trade unionists, who are aware that news production is so routinised that an early appearance can ensure their presence in all of the day's bulletins.

The variations we have indicated are on a small scale which affects neither the bulletin formats nor the general distribution of content categories. They simply indicate that there is a compromise between the demand for immediacy and up-to-dateness and the constraints of preparing a well illustrated bulletin (i.e. increasing use of film). This provides some evidence that gives the lie to the received notion that the television offers 'the news as it happens'; although in certain areas of political life this may be increasingly true as the events themselves (speeches, news conferences, statements, etc.) become media oriented.

Weekend variations

The pattern of events as well as the organisation of news gathering changes at weekends in certain predictable ways. It is apparent from the percentage distribution of items by channel shown in Table 4.7 that there are some very clear differences between weekend and weekday bulletins. There is once again close similarity between channels in most respects. The majority of categories tend to have relatively fewer items at weekends – this is true of the political, economic, home affairs, disasters and science categories on all three channels. Although major sporting events occur most frequently at weekends, there is no general increase in the percentage of items devoted to sport, since this is widely covered in specialist programmes. Major political speeches are often made at weekends and are

Table 4.7 Distribution of items by channel, weekdays and weekends

	BBC1 Week-days	Week-ends	BBC2 Week-days	Week-ends	ITN Week-days	Week-ends
10 Political	21·6	17·4	24·5	18·0	22·1	18·0
20 Industrial	16·8	25·0	16·9	19·4	14·4	23·3
30 Foreign	22·0	27·0	23·3	34·6	25·7	31·2
40 Economic	8·5	3·6	8·3	4·2	7·5	3·1
50 Crime	4·6	6·4	4·1	3·9	5·0	4·8
60 Home Affairs	6·5	4·0	6·3	5·5	6·3	5·0
70 Sport	7·5	5·9	4·5	4·2	6·8	7·2
80 Human Interest	5·2	3·8	4·7	6·1	5·3	3·1
90 Disasters	6·8	6·2	6·1	3·9	5·4	3·9
00 Science	0·9	0·6	1·3	0·3	1·6	0·4
Total %	99·9	99·9	100·0	100·1	100·1	100·0
n =	1835	471	774	361	2017	484

reported, but this does not offset the general decline in political coverage on these days. All this must be related to the reduced manpower on duty in the newsrooms. The only categories which are proportionally much bigger on each channel are the industrial and foreign categories, and in each case the difference is quite substantial.

This average increase of 6·5 per cent for industrial and 7·3 per cent for foreign items is attributable to much foreign news being available via the EBU, partly processed, as it were. This makes it more likely to be included at weekends when the domestic resources of the newsroom are limited.

The higher proportion of industrial news is more difficult to account for. It is unlikely to reflect a greater amount of activity in the industrial sphere. However, much industrial coverage is 'crisis' news of a kind which can still be carried on a 'thin' day; trade union conferences often occur at weekends; and our classification system placed many political speeches directly on industrial questions in the industrial category (e.g. Michael Foot's major confrontation with steel workers at Ebbw Vale). Whatever the changes in profile at weekends, there are only minor differences in presentation, use of film, and so on. There is a change in the ratio of items with voice-over as against in-vision film (see below p. 125) which reflects reduced manpower in the field. But the overall use of film remains constant as the decrease in in-vision film is offset by an increase in the use of voice-over film. This generalisation does not apply to *News Review*, which is adapted to a specific audience (the deaf) and as a 'review' chooses from all the stories of the preceding week.

The variations so far referred to in this section have been fairly superficial. They do not alter in any significant way the formats and priorities described earlier. What variations there are could be anticipated. These relate in essence to the diminished activity of news gatherers, world-wide, at weekends.

Variations from week to week

Figures 4.7, 4.8 and 4.9 show the variation in the time devoted to each content category by channel from week to week for the first twelve weeks of 1975. The first and most obvious feature to be noticed is the close similarity of profile between the three channels. The relative weighting of categories and the directions of variations follow the same pattern. This is of course to be expected, given the correspondences between structural and professional norms already discussed. The three major categories of political, industrial and foreign never occupy less than 50 per cent of total news time in any one week and usually account for much more. Conversely it is rare for any single other category to take up more than 10 per cent of the time in a week. A second feature to be noticed from Figures 4.7–4.9 is the obvious emphasis on political news in January and the first week in February and the build-up of foreign news from the middle of February until the end of the sample period. The proportion of industrial news does not show any such trend, although there is a marked drop on all three channels in the week beginning 27 January.

As would be expected from the low standard deviation in the size of the sports and human interest categories, they fluctuate within very narrow limits and do not show any long-term tendency to increase or decrease. These profiles are the product of a compromise between prearranged bulletin structures – the result of routine processing practices and the use of a limited interpretive framework – and the untidy, erratic patterns of domestic and world events.

Politics

The three main peaks of political coverage occurred in the weeks beginning 3 and 24 February and 17 March, on the occasions of the Conservative leadership struggle, the publication of the EEC Referendum proposals and the arrest of John Stonehouse MP. Only in the first of these can the story be said to have swamped the rest of the news and then only on the day of Mrs Thatcher's victory.

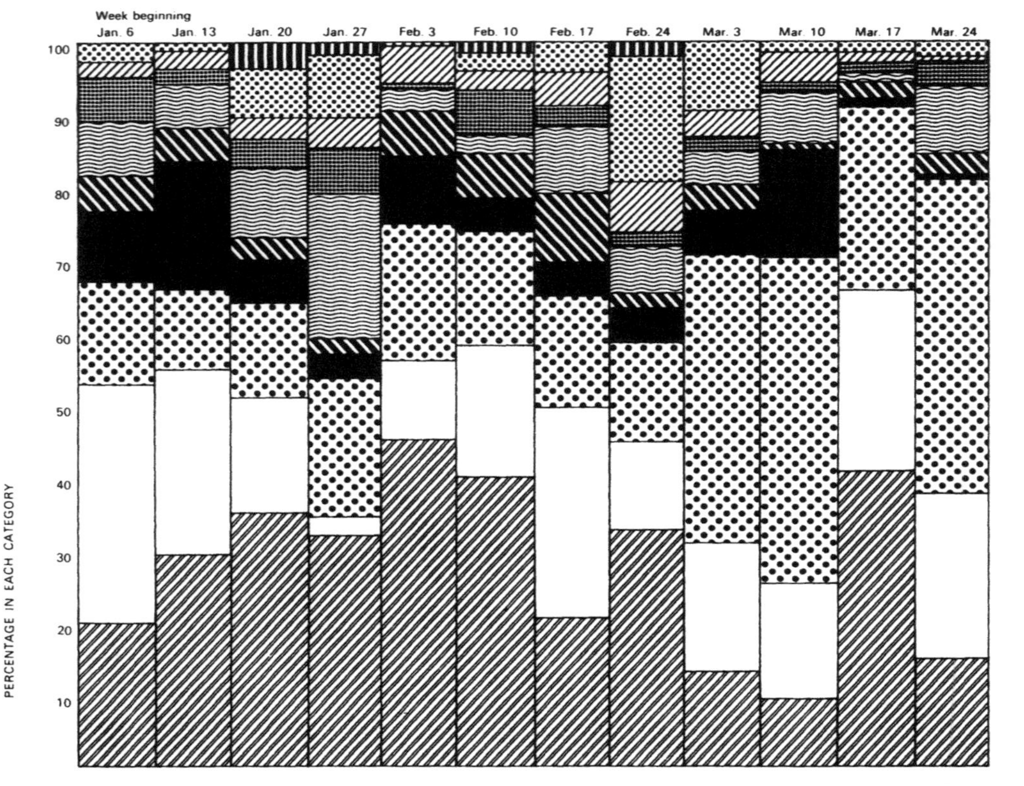

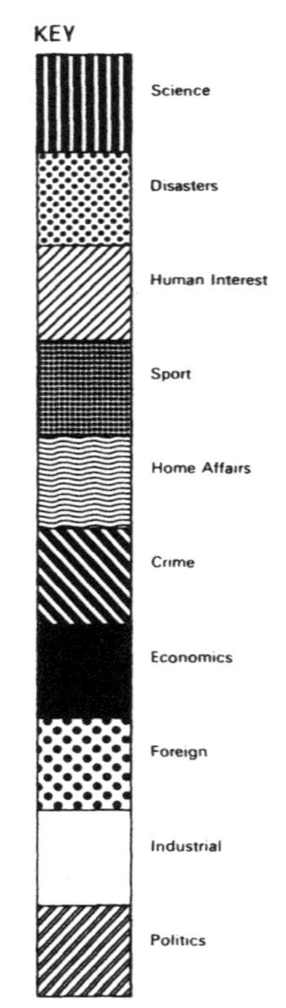

Figure 4.7 Longitudinal variation in story categories — BBC1 (BBC1 weekday bulletins, January–March 1975)

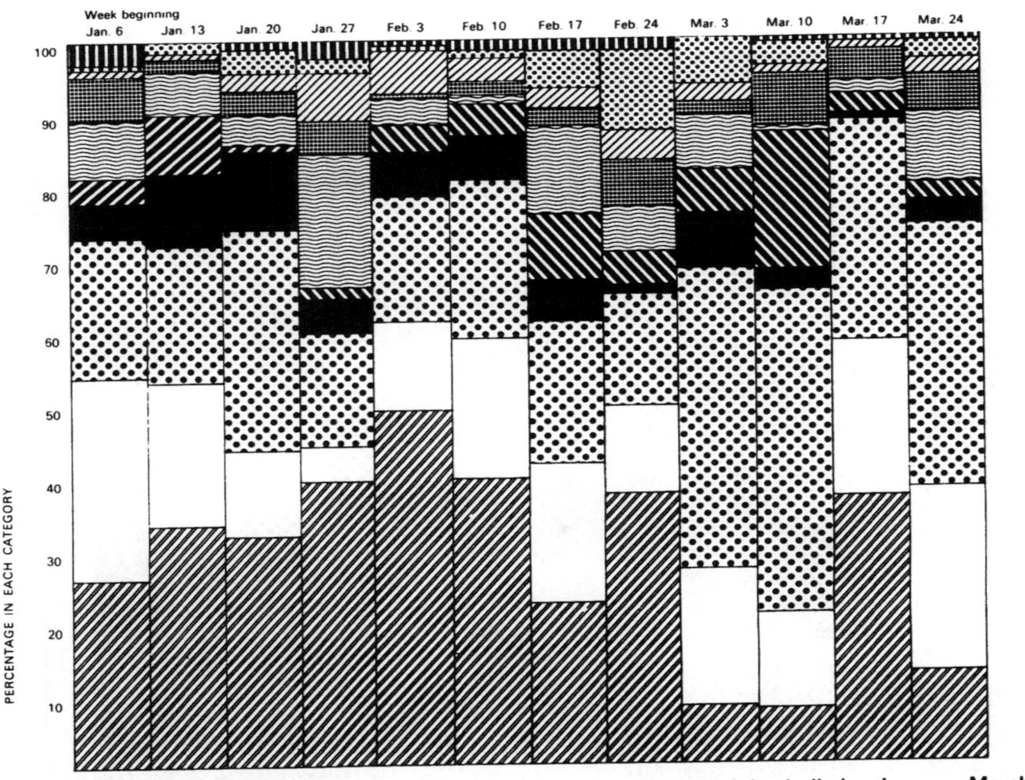

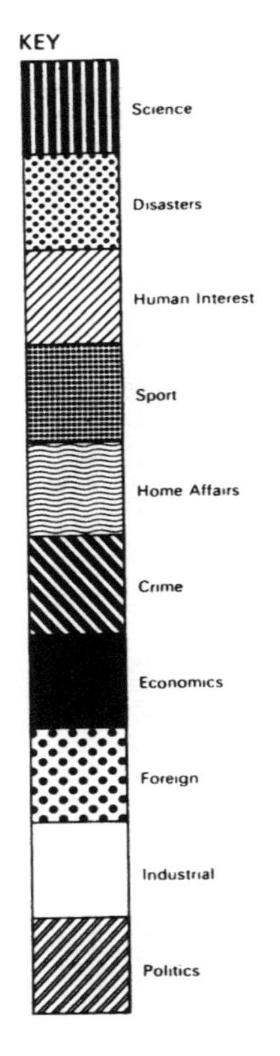

Figure 4.8 Longitudinal variation in story categories — ITN (ITN weekday bulletins January–March 1975)

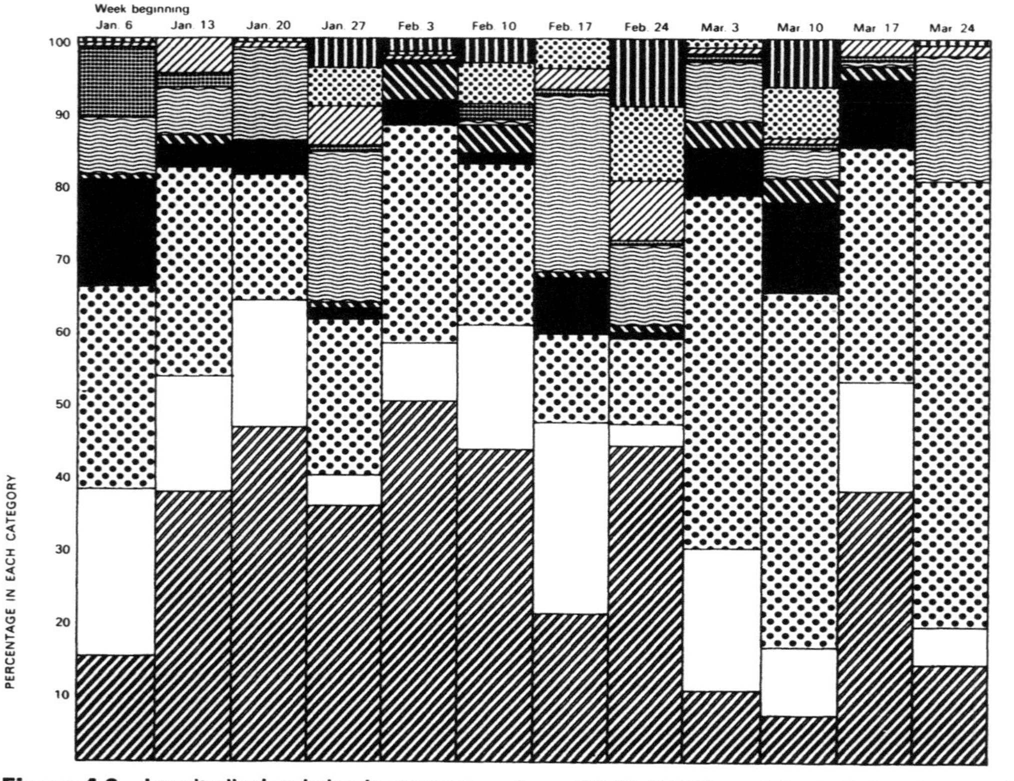

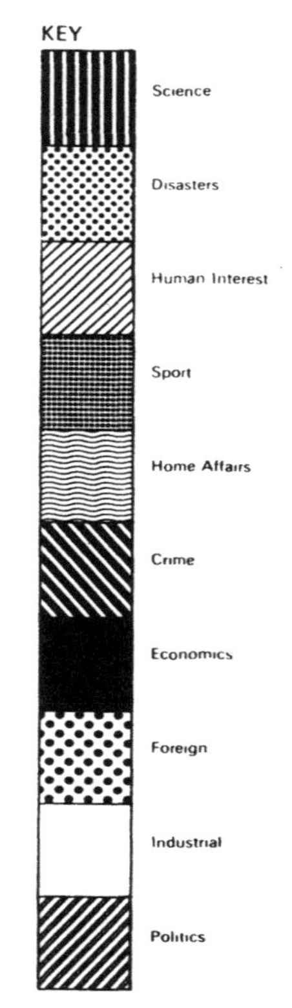

Figure 4.9 Longitudinal variation in story categories — BBC2 (BBC2 weekday bulletins, January–March 1975)

Thus the amount of news in any story category is unlikely to be accounted for by a single major story. The Conservative leadership struggle coincided with heavy Northern Ireland coverage and the Guildford pub bombs trial; the Referendum issue occurred at the same time as the Manchester bombs trial and the Civil List debate; and the Stonehouse story occurred at a time when the threatened split over the EEC in the Labour Party was being reported. Despite the fairly complex combination of stories and subject matter in the political category, the peaks and troughs of the profiles on all three channels are perfectly parallel. But it could be noted, in general, that in the week that Hué fell and the South Vietnamese retreat became a rout, coverage of South East Asia hardly increased and more time was spent on domestic politics, including the arrest of John Stonehouse.

Foreign

The increasing pace of events in Vietnam and Cambodia in part account for the substantial increase of foreign coverage towards the end of the sample period. Another major running story at this time was Portugal, which included an attempted coup. The tendency to give lengthy coverage is particularly strong on BBC2's *News Extra.* Bearing in mind that standard deviations showed foreign news to be more flexible than most other categories, the average duration of foreign news is not a useful measure.

Industrial

This is the most stable of the three major categories on all three channels. Only in the first week did the proportion of industrial news rise above 30 per cent (and then only on BBC1), this being due to a range of stories (from the car industry, the health service and elsewhere) rather than to one major story. Predictably, the proportion of industrial news on BBC2 fluctuates within narrower limits than on the other two channels although the trend remains quite similar. Also, instead of there being slightly more industrial news towards the end of the period, this was not so on BBC2; there, foreign coverage was proportionally increased at the expense, as it were, of industrial. This lack of emphasis on industrial news on this channel is not unexpected, given the content profiles discussed earlier. A detailed discussion of the stories which comprise the industrial categories and comparison with independently derived indices of activity will be found in Chapter 5.

Other categories

The only notable exception to the general pattern of a finely-ground selection of stories in the other story categories, occupying about 40 per cent of the total news time, is the very small allocation of these categories in the last fortnight in March. This is exceptional in terms of the profiles shown in Figures 4.7–4.9. It is interesting to note that in this period the reverse was not true: there was no week in which, for example, disaster, economic or home affairs news – or a combination of these – took up a majority of the news time. The weeks in which outstanding events did occur – the Moorgate Tube disaster in the week beginning 24 February, for example, swamped one day's coverage – did little to alter this general picture. The second week in January, which saw the publication of several sets of figures on trade, prices and business activity significantly boosting the size of the economic category, still had less than 40 per cent of news in these remaining categories. It is unlikely, with current professional assumptions about news values, that the pattern can ever change in any major aspect from the one shown in these profiles. In other words, whatever happens in the real world, we would suggest, will be accommodated in the pattern presented here. This pattern is the essential of the coding of the bulletin structures. It is the surface expression of the interpretive frameworks used by television news personnel.

Placement

The arrangements of news items to form a programme rather than a haphazard catalogue of events is one of the most characteristic features of television news and is no less important than the layout of a newspaper. The design of a newspaper is intended to catch the reader's eye and lead it in certain directions. So in television association of subject matter, expressed most often by verbal links, provides important clues to the intentions of news producers. Placement, and the juxtapositioning this entails, reveals the nature of the programme as something more than the sum of its individual items. Within certain limits the placing of items in relation to one another by content categories is predictable. The categories from which lead stories are most frequently chosen are political and industrial, with an almost equal probability of foreign news. This can be seen from Tables 4.8–4.11 which present in detail the proportion of stories which fall into each item placing (lead, payoff, etc.) in the case of two lengths of programme – all nine-item and all twelve-item news programmes on BBC1 and ITN in the three-month period. We shall call this the item ranking. The patterns described below apply to nine- and twelve-item bulletins whenever they were transmitted – most typically BBC1's Early Bulletin is nine items and ITN's *News at Ten* is twelve. But the points made below would apply to the BBC's main *Nine O'Clock News* or ITN's Early Bulletin if they were to contain nine items.

In bulletins with nine items (see Tables 4.8 and 4.9) there are certain obvious and predictable features in the placing of stories. There is not only a stronger tendency to lead with items in the larger categories of politics, industrial and foreign than would be expected from a random distribution but also a complete absence of lead stories in the home affairs, sport, human interest and science categories. This applies to both BBC1 and ITN. The closing items show a similar pattern in reverse. Bulletins rarely conclude with industrial, foreign or political stories. Within this largely taken for granted rank order are contained less obvious but equally important regularities – important that is, for audience response, since primacy is directly related to the capacity for comprehension, cognitive evaluation and recall.[15] Stories in the political categories are generally well distributed throughout the rank order although they nevertheless take up a larger proportion of lead items on BBC1 and ITN respectively. A comparison between the separate political categories, however, shows that there is a significant difference between the 27 BBC1 bulletins represented in Table 4.8 and the 19 ITN bulletins represented in Table 4.9. ITN carried a higher proportion of Northern Ireland stories and relatively fewer 'other political' stories (categories 14–19) than item one; parliamentary news was also given less priority. This is largely accounted for by the higher priority given by ITN to disaster and crime stories. It is to this extent that ITN can be said to be more 'popular' than BBC. The industrial category, although one of the largest categories on both channels, tends to be used less often as a source for lead items than either political or foreign news. However, this does not mean that industrial stories are evenly distributed throughout the bulletin. The tables show that on both channels they cluster in the first half of the programme and

Table 4.8 Distribution of story placings in nine-item programmes (BBC1). Figures are percentages of row totals.

Rank	Category: Parliamentary	N. Ireland	Terrorist	Other Political	Industrial	UK and World	Foreign	Economics	Crime	Home Affairs	Sport	Human Interest	Disasters	Science	Row %
Item 1	11·1	14·8	—	22·2	14·8	3·7	18·5	3·7	3·7	—	—	—	7·4	—	100
Item 2	—	3·7	3·7	3·7	40·7	3·7	14·8	7·2	—	—	—	3·7	14·8	3·7	100
Item 3	3·7	11·1	—	7·4	18·5	7·4	11·1	25·9	—	3·7	—	3·7	7·4	—	100
Item 4	—	11·1	—	7·4	25·9	14·8	33·3	—	3·7	—	—	—	—	3·7	100
Item 5	3·7	—	—	14·8	14·8	3·7	33·3	14·8	—	3·7	—	3·7	7·4	—	100
Item 6	3·7	11·1	—	22·0	3·7	3·7	25·9	3·7	11·1	—	3·7	7·4	3·7	—	100
Item 7	—	11·1	—	7·4	11·1	3·7	11·1	7·2	3·7	11·1	11·1	11·1	11·1	—	100
Item 8	7·7	—	—	3·7	7·7	—	3·8	—	19·2	7·7	26·9	15·4	7·7	—	100
Item 9	—	3·7	—	3·7	3·7	—	7·4	—	—	18·5	29·6	25·9	7·4	—	100
n =	8	18	1	25	38	11	43	17	11	12	19	19	18	2	Total 242

Table 4.9 Distribution of story placings in nine-item programmes (ITN). Figures are percentages of row totals.

	Category														
Rank	Parliamentary	N. Ireland	Terrorist	Other Political	Industrial	UK and World	Foreign	Economics	Crime	Home Affairs	Sport	Human Interest	Disasters	Science	Row %
Item 1	—	21·1	10·5	5·3	10·5	5·3	15·8	5·3	10·5	—	—	—	15·8	—	100
Item 2	—	10·5	5·3	21·1	21·1	—	15·8	15·8	5·3	5·3	—	—	—	—	100
Item 3	—	15·8	5·3	5·3	31·6	15·8	—	21·1	—	—	—	—	5·3	—	100
Item 4	5·3	10·5	—	5·3	26·3	10·5	15·8	15·8	—	—	—	—	5·3	—	100
Item 5	10·5	5·3	—	10·6	5·3	5·3	42·1	5·3	5·3	5·3	—	—	5·3	—	100
Item 6	5·3	10·5	5·3	—	15·8	15·8	42·1	—	—	5·3	—	—	—	—	100
Item 7	—	5·3	—	10·5	21·1	5·3	36·8	—	—	5·3	5·3	5·3	—	5·3	100
Item 8	5·3	5·3	—	—	—	—	15·8	5·3	5·3	5·3	26·3	26·3	5·3	—	100
Item 9	—	—	—	—	—	10·5	5·3	—	—	15·8	26·3	21·1	21·1	—	100
															Total
n =	5	16	5	11	25	13	36	13	5	8	11	10	11	2	171

Table 4.10 Distribution of story placings in twelve-item programmes (BBC1). Figures are percentages of row totals.

	Category														
Rank	Parliamentary	N. Ireland	Terrorist	Other Political	Industrial	UK and World	Foreign	Economics	Crime	Home Affairs	Sport	Human Interest	Disasters	Science	Row %
Item 1	11·5	7·7	11·5	3·8	11·5	7·7	11·5	3·8	19·2	3·8	—	—	7·7	—	100
Item 2	—	7·7	7·7	15·3	23·1	7·7	15·4	15·4	—	—	3·8	—	3·8	—	100
Item 3	—	7·7	11·5	7·7	30·8	15·4	7·7	15·3	3·8	—	—	—	—	—	100
Item 4	—	3·8	—	11·5	34·6	3·8	11·5	11·5	3·8	11·5	—	3·8	3·8	—	100
Item 5	11·5	3·8	7·7	3·8	26·9	—	7·7	23·1	—	3·8	3·8	—	7·7	—	100
Item 6	3·8	3·8	3·8	7·7	34·6	—	7·7	7·7	7·7	—	3·8	7·7	11·5	—	100
Item 7	—	3·8	11·5	3·8	15·4	11·5	30·8	7·7	11·5	—	—	—	3·8	—	100
Item 8	3·8	7·7	3·8	7·7	19·2	3·8	26·9	11·5	7·7	—	—	3·8	3·8	—	100
Item 9	—	3·8	3·8	11·5	11·5	23·1	23·1	3·8	3·8	—	—	—	15·4	—	100
Item 10	—	—	3·8	11·5	19·2	—	30·8	—	7·7	—	3·8	7·7	3·8	11·5	100
Item 11	3·8	7·7	—	15·4	7·7	7·7	3·8	—	3·8	7·7	11·5	3·8	11·5	15·4	100
Item 12	—	—	7·7	—	—	—	11·5	—	—	7·7	65·4	7·7	—	—	100
n =	9	15	19	26	61	21	49	26	18	9	24	9	19	7	Total 312

Table 4.11 Distribution of story placings in twelve-item programmes (ITN). Figures are percentages of row totals.

Rank	Category														
	Parliamentary	N. Ireland	Terrorist	Other Political	Industrial	UK and World	Foreign	Economics	Crime	Home Affairs	Sport	Human Interest	Disasters	Science	Row %
Item 1	10·5	10·5	5·3	5·3	26·3	5·3	15·8	10·6	—	—	—	—	10·5	—	100
Item 2	15·8	5·3	—	21·1	21·1	—	31·6	—	—	—	—	—	5·3	—	100
Item 3	15·8	—	10·5	10·5	31·6	5·3	21·1	—	5·3	—	—	—	—	—	100
Item 4	5·3	10·5	5·3	5·3	21·1	—	26·3	5·3	5·3	5·3	—	—	10·5	—	100
Item 5	10·5	10·5	5·3	5·3	10·5	—	10·5	31·6	15·8	—	—	—	—	—	100
Item 6	—	—	—	5·3	26·3	15·8	21·1	10·5	5·3	—	—	5·3	10·5	—	100
Item 7	—	5·3	5·3	5·3	21·1	15·8	26·3	—	10·5	—	—	5·3	5·3	—	100
Item 8	5·0	15·0	5·0	—	15·0	5·0	35·0	—	5·0	5·0	5·0	—	5·0	—	100
Item 9	5·6	5·6	—	5·6	—	22·2	33·3	11·2	—	5·6	5·6	—	5·6	—	100
Item 10	—	—	—	10·5	—	10·5	36·8	10·6	—	5·3	10·5	5·3	5·3	5·3	100
Item 11	—	—	5·3	—	10·5	5·3	10·5	5·3	—	15·8	31·6	10·5	—	5·3	100
Item 12	—	—	—	—	—	—	10·5	—	5·3	26·4	31·6	26·3	—	—	100
n =	13	12	8	14	35	16	53	16	10	12	16	10	11	2	Total 228

occupy the largest proportion (41 per cent) of second items on BBC1 and the largest proportion of third items on ITN. This feature of story placings in which industrial takes third place to political and foreign news in the lead item but displaces them in the second and third items may be interpreted as a product of decisions based on two partly conflicting criteria. Since industrial news often embodies high news values (large numbers of people involved or affected, for example) simultaneously with a reputation for tedium, complexity and difficulty of presentation, it is not surprising to see it in a consistently prominent, but not the most prominent, position.

The implication of this juxtapositioning (especially between political, industrial and economic stories) may be more significant than is usually allowed for. Analysis of the structure of the discourse and the meaning of verbal connections between items is necessary to give substance to this in terms of the language of the news.

A study of this will be included in a forthcoming volume. Consider the following example by way of illustration. An item announcing the Department of Employment's unemployment figures in January 1975, which showed a significant increase in the number out of work, was immediately followed by the phrase 'Meanwhile, the rise in share prices continues. . .' (BBC2, 19.30, 23 January 1975). The verbal link suggests a degree of association which is in fact denied by the substance of the item. The formal connection in discourse is an expression of the artificiality of the programme structure; the denial of any substantial connection shows the dominance of the logic of fragmentation. But the continuous use of subject matter in a preferred order must in itself provide the elements of a structure of interpretation, aside from language. If the dominant logic of the news is to present fragments of information with no general coherence apart from that of the bulletin structure, this structure cannot be dissociated from a taken-for-granted interpretation of the world above and beyond the 'facts' and 'events' being reported. It cannot simply be justified in terms of professional criteria of reporting and presentation; the placing of stories demonstrated very clearly that this is a structure which mediates in quite specific ways the information which is transmitted. Several inferences can be made from Tables 4.8 and 4.9. The close proximity between economic and industrial items, which is particularly clear on BBC1, suggests that items about particular industrial situations are likely to be juxtaposed with items (usually shorter) on the general state of the economy, with a resultant strong implication of a causal connection.

The bulk of foreign items are placed in the middle of the programme – items 4–6 on BBC1 and items 5–7 on ITN. Domestic news more often than not has priority and for purpose of presentation, foreign stories tend to be grouped together because the foreign/domestic distinction is the most fundamental of all. The logic of this distinction, which is signalled by phrases such as 'And now back home . . .' or 'Now some foreign news . . .', can override possible alternative criteria of placement.

One of the few significant between-channel differences is the lower prominence given to crime news on BBC1 compared with ITN. Also the tendency to place disaster items either at the beginning or at the end of a programme (see above p. 103) is more marked on ITN than on BBC1. In general, however, the similarity between channels which we have noted in other contexts is to be found here as well.

Comparable data for programmes with fewer than nine items (not reproduced here) do not show any major departures from the patterns of Tables 4.8 and 4.9. The relative priority of political, industrial and foreign news remains the same, as does the position of sports and human interest items. The tendency for disaster items to be either at the beginning or at the end of the programme is no less marked. Home affairs, economic and crime stories do appear to be more randomly distributed but no firm conclusions can be based on the small number of items of these categories in these shorter bulletins.

One difference between the nine-item and twelve-item programmes is that the latter have a slightly more evenly distributed pattern. Tables 4.10 and 4.11 show that although the general priorities remain the same, lead stories are taken from a larger proportion of categories, with political stories rather less dominant than they are in the nine-item bulletins, described above. The longer the programme, the greater the flexibility of presentation – at least in theory. So these results in practice are not unexpected. The effect of the commercial break on *News at Ten* is not directly apparent from these tables, since other programmes are included as well. However, there is no evidence of any major differences in priorities or story placings because of this. A comparison between ITN and BBC1 shows somewhat more prominence given to industrial news on the former, i.e. a higher percentage of lead stories are industrial on ITN. The difference is not very marked and the relationships between political, industrial and economic items described for the nine-item programmes remain the same.[16]

Each bulletin represents a selection from a greater number of processed news stories than are eventually run. There are items in the running order which are dropped, cut or exchanged for reasons of overall bulletin timing. Whenever there is important news to be updated in the course of a programme this will distort the generally predictable pattern.

In the analysis presented so far, we have implied that the 'artificiality' of the news, which is generally hidden from the consumer, is clearly seen in the shape of bulletin profiles and especially in the limited ways in which they vary. The news viewer, who is used to identifying and experiencing the news as a sequence of unconnected stories, does not normally reflect on these structural features. They appear as 'natural'. There is no directly available alternative outside the narrow range of profiles which we have demonstrated. Yet a comparison with the press and with alternative descriptions of the world contained in official statistics will show that alternative priorities and treatments are conceivable.[17]

The practices which lead to highly predictable bulletin profiles involve allocations of time and resources which have significance apart from the content particular to each story. The ways in which news is regularly shaped and processed for television partly determines what can be said. These practices have the characteristics of a *code*: as well as being a means to communicate, they are a social index of a system of values. Just as the language of the news – its vocabulary, grammar and style – contains a meaning, so too the features we have described are *carriers* of meaning. The rather stable proportion of total television output which is taken up by news bulletins is one of the chief defining characteristics of news. It is that which can be itemised and contained within a bulletin. Within this narrow context, 'developments' and 'processes' must always be immobilised and presented in the news in terms of crisis, novelty or as an important departure from the normal course of events. Thus the rules by which newsworthy events are identified are rules for transforming these events into a new kind of product. Once removed from the circumstances in which it originates, news is only recognisable as an expression of newsmen's calculated and orderly division of labour and division of the world. The regular distribution of events in the categories we have described derives from the need for a constant output of news in areas which can be anticipated rather than from long-term movements, poorly organised groups, or remote locations which defy the routines of news gathering. In a similar way the placing of stories within each bulletin is evidence of a normative ordering. So far we have only demonstrated that this ordering exists and that it embodies certain values. These are not only 'news values' in the professional journalistic sense but also social values which amount to a ritual re-enactment of the basis of order in society.

Technical inputs

For a number of reasons, television's potential as a visual medium is exploited in the news in ways which vary from story category to story category and to a lesser extent from channel to channel. The news as historically determined is an essentially verbal written activity, and remains so on television which relies on news personnel (casters, presenters, correspondents and reporters) to read this largely written news to the audience. To avoid what is professionally regarded in other television production areas as the potential tyranny and monotony of the 'talking head', that is close-ups of faces talking, a great deal of effort is expended to make the bulletins visually interesting. We will argue in the next volume that these visual inputs are most often subordinate to the text and used illustratively. In the conflict between the codes of journalism and film, a conflict which lies at the heart of the bulletins, journalism basically triumphs. Only rarely do film or other visual inputs from outside the studio get into the bulletins because of their own particular quality – such as when the material is exclusive, exceptionally dramatic and has unusual immediacy. The BBC's use of air-to-air footage of a hang-gliding pilot, the general use of the Birmingham bomb aftermath film (rushed unedited to the screen) and the 'last plane out of Da Nang', are examples of this. Tables 4.12, 4.13 and 4.14 therefore do not mainly refer to this type of material. However, although subordinate to the structures already described, the inputs here represent nevertheless an important level of mediation.

Film

Film, either with or without commentary (F/X as against SOF in our notation), is by far the most important input. This division of film material enables us to isolate at least crudely, the non *talking head* elements in the bulletins. This is crude because within the SOF category, as we used the term,[18] no distinction was made

Table 4.12 Technical inputs by category (BBC1, all bulletins). Figures show the percentage of items in each category which contain the given input and are rounded to the nearest 1 per cent.

Input	Category: Parliamentary	N. Ireland	Terrorist	Other Political	Industrial	UK and World	Foreign	Economics	Crime	Home Affairs	Sport	Human Interest	Disasters	Science	Average (all categories)
SOF	17	23	33	40	31	21	20	20	33	37	12	29	20	30	26
F/X	19	49	55	38	39	38	53	21	42	43	32	59	52	35	41
BW	—	1	3	2	1	2	7	—*	2	1	—	8	3	10	3
POOL A	—	—	—	—	—	1	3	—	—	—	1	—	1	—	—*
LIVE	7	4	1	2	1	4	—	2	1	3	1	—	2	—	2
VTR	14	7	6	13	8	11	5	3	7	9	15	4	5	20	9
ST	11	8	4	8	4	5	1	1	5	4	—	—	2	5	4
REM	2	4	1	2	1	7	1	1	2	4	—	—	—	5	2
OB	1	—	—	4	1	1	2	—	—	3	13	3	3	10	3
EBU	—	1	1	—	—	5	4	2	—	—	2	2	3	10	2
SAT	—	—	—	3	—	1	4	—	1	—	6	1	1	—	1
POOL B	—	—	—	1	—	1	2	—	1	—	1	2	—	10	1
LIB	2	3	—	5	2	2	4	3	5	6	2	9	1	10	4
RC	4	1	3	—	—	11	7	6	3	1	1	2	1	—	3
n =	86	139	69	185	427	168	363	173	114	138	165	114	145	20	Total 2306

* < 0·5 per cent.

Table 4.13 Technical inputs by category (BBC2, all bulletins). Figures show the percentage of items in each category which contain the given input and are rounded to the nearest 1 per cent.

	Category														
Input	Parliamentary	N. Ireland	Terrorist	Other Political	Industrial	UK and World	Foreign	Economics	Crime	Home Affairs	Sport	Human Interest	Disasters	Science	Average (all categories)
SOF	20	23	21	31	21	24	18	13	20	28	8	29	28	55	24
F/X	22	34	29	24	30	30	46	17	22	38	50	55	43	64	36
BW	—	1	—	1	1	2	3	—	—	—	2	5	2	18	3
POOL A	—	—	—	—	—	3	2	—	—	—	2	—	2	—	1
LIVE	4	1	4	2	2	3	1	—	2	7	—	—	—	—	2
VTR	4	9	—	10	6	11	7	6	—	19	22	2	5	27	9
ST	2	8	—	6	3	2	1	3	—	2	2	—	—	—	2
REM	—	3	—	5	1	9	—	1	2	1	—	—	—	—	2
OB	—	—	4	2	1	2	1	—	—	15	18	2	3	27	5
EBU	—	—	—	—	—	7	3	—	—	—	—	—	3	18	2
SAT	—	—	—	2	1	2	5	—	—	—	4	2	2	—	1
POOL B	—	—	—	1	—	3	2	—	—	—	4	—	2	18	2
LIB	2	5	4	2	4	2	5	3	2	5	2	7	—	18	4
RC	2	1	—	—	1	9	5	—	2	2	—	2	3	—	2
n =	54	77	24	100	201	93	212	79	46	69	50	58	61	11	Total 1135

Table 4.14 Technical inputs by category (ITN, all bulletins). Figures show the percentage of items in each category which contain the given input and are rounded to the nearest 1 per cent

	Category														
Input	Parliamentary	N. Ireland	Terrorist	Other Political	Industrial	UK and World	Foreign	Economics	Crime	Home Affairs	Sport	Human Interest	Disasters	Science	Average (all categories)
SOF	14	15	32	20	26	23	10	22	25	38	15	30	15	29	22
F/X	12	50	54	38	43	43	38	23	38	46	36	64	49	41	41
BW	—	—	—	2	1	3	5	—	2	1	1	3	3	6	2
POOL A	—	1	—	—	—	1	3	—	—	—	—	2	2	—	1
LIVE	15	4	4	6	6	5	3	2	4	6	2	1	2	—	4
VTR	12	4	6	16	6	9	5	2	4	6	6	2	7	9	7
ST	6	5	—	4	4	3	1	—	3	3	—	—	1	—	2
REM	2	2	1	1	—	5	—	—*	—	2	—	—	—	—	1
OB	2	—	4	6	2	1	—	—	1	—	4	1	6	6	2
EBU	—	—	3	—	—	5	3	—	—	—	2	1	2	9	2
SAT	—	—	—	1	—	1	3	—	—	1	2	—	2	—	1
POOL B	—	—	—	—*	—	2	2	—	—	—	1	1	1	6	1
LIB	3	3	5	8	3	3	5	4	7	3	—	10	6	15	5
RC	—	4	4	3	1	5	3	1	3	—	2	—	2	—	2
n =	111	161	78	182	404	190	480	167	123	151	172	121	127	34	Total 2501

* < 0.5 per cent

between the *talking head* and other shots in the film. Thus the distinction between SOF and F/X as we have drawn it underestimates the non *talking head* film element in the bulletin. More than one item in three makes use of at least one piece of film.[19]

The overall difference between channels is not very great, particularly in the case of BBC1 and ITN. The slightly lower figure for voice-over film on BBC2 can be attributed to the effect of the short *Newsday* summary which rarely allows time for the transmission of film and the fact that the longer average item duration on *News Extra* allows most items to have multiple inputs, which are not measured here. The ratio of film with *talking heads* to voice-over film is slightly higher on BBC than on ITN. This is because BBC have a greater propensity to allow their reporters to appear in vision than ITN.

A breakdown of these results by category demonstrates that within this general similarity the inputs are far from evenly distributed. First, certain categories have a higher level of both kinds of film input than others have; journalistic requirements in some categories more totally override the programmes' general 'need' for film. Second, the ratio between in-vision film and voice-over film varies between categories in predictable ways. The categories which have the highest level of voice-over inputs on both BBC1 and ITN are human interest and disasters. The hypothesis is that human interest and disasters will survive the process of selection better if attached to good, i.e. visually interesting – not *talking head* film. This is the most essentially 'televisual' element of the bulletins. Good film in this sense is exactly what no other medium can so immediately bring before its audience. The use of such film is an essential part of the 'news as it happens' notion of the bulletins. It is interesting to note that across all bulletins, 61 per cent of material is not illustrated in this way – it does contain *talking heads*. When the routine nature of much voice-over newsfilm is also taken into account, it becomes possible to claim that the news is, by professional criteria, visually extremely unexciting. Basically, despite still photographs, graphics and the rest, most of the time it is *talking heads*, in studio or on location.

Foreign and home affairs news, like human interest and disasters, have above-average inputs of these types for the same reasons as noted above. Political, industrial and crime stories have about average inputs while the story categories with relatively little film are sport and economics (both of which of course make frequent use of graphics to convey scores and statistics). Further, a certain amount of non-results sports coverage uses videotapes generated by sports departments. BBC2 differs from this pattern only slightly; both science and sports items are more likely to use film than on the other two channels. Here again we would suggest that this is due to *News Extra*'s current affairs approach to the bulletin format with the attendant increase in the number of feature items.

We have indicated above that in so far as people appear in vision, they contribute to the preponderance of *talking heads* in the bulletins. But we have pointed out that the SOF category did not distinguish between in-vision and voice-over. Once a person had appeared in-vision, the whole piece was so categorised. And further to this, there is a distinction between in-vision reporting and interviews on film and newscaster/correspondent studio pieces. On film, the television personnel and their interviewees give the impression of immediate on-the-spot reporting. Even reporters to camera in-vision pieces on film, although often scripted, do not go through the script sub-editing and dubbing processes used in voice-over commentary. Thus the higher the percentage of in-vision film, the greater the proportion of participants (including television personnel) who speak more on their own behalf as identifiable individuals rather than in the more anonymous mode of the studio script. The ratio between in-vision and voice-over film is therefore a measure, albeit a crude one, of the dominance of personally attributed speech in settings other than the studio. The tables reveal some major variations which are largely consistent across channels. Some of the most notable are to be found in the political categories. In the parliamentary category there is an almost equal amount of in-vision and voice-over film (i.e. a relatively high ratio) which is accounted for by the high proportion of filmed interviews with political figures. But it is important to note, for instance, that the corresponding ratio for Northern Ireland coverage is very low (23:49 per cent on BBC1 and 15:50 per cent on ITN) except on BBC2. In the film coming from Northern Ireland, there is a substantial element of 'aftermath' and other dramatic voice-over coverage. The same is true of the terrorist story category but to a lesser degree. One side effect of this could be that Northern Ireland stories are perceived as more highly mediated, with less identifiable individual comment, than is the case with Westminster politics.

For industrial stories the ratio between the two film types we have distinguished is slightly above average on all three channels. This could be a result of the 'events' orientation we have argued exists in industrial coverage. In seeking 'photogenic discord', the newsrooms, especially ITN with a voice-over/in-vision ratio of 43 to 26 per cent, use industry as an area with more limited *talking heads.* Foreign stories have the lowest ratio of all, obviously because of the convention that subtitles are almost never used. Foreign correspondents are expensive and much film arrives from non-English speaking networks without *talking heads.* This lack is partly made up for by the use of radio links (RC) in about 7 per cent of foreign items. Other story categories with low ratios of in-vision to voice-over include sports, disasters and, for less obvious reasons, human interest stories, including obituary items. Relatively high ratios are found in the economics, home affairs and science story categories which regularly contain interviews with experts and spokesmen.

Black and white film

The use of black and white film is now limited to a small range of circumstances. As we have said above, it is of foreign origin or is from the outermost points of the British network or is old film from the library used to provide background. Therefore, the concentration of its use in the foreign, politicians (personal), human interest and science categories is to be expected. Like some other of the technical input categories noted on the initial logging sheet, this we would now argue is not of much significance.

Library film

The use of library film is quite evenly spread throughout the range of categories. The most frequent uses are in obituaries, 'political obituaries' (as we might describe the coverage of Mr Heath's fall from office), science stories and background to reports just published – accident inquiries, commissions, etc. Library film is not generally used to elucidate events which are still newsworthy in their own right. Arguably it does not even have this function in the cases mentioned; it simply provides an occasion for the use of rare, dramatic, entertaining and relevant pieces of film. A further use of library shots within up-to-the-minute film reports was almost impossible to note. Virtually the only exception to this is the one mentioned in Appendix 2 – black and white coverage of developments in the North Sea Oil story cut with colour footage of oil rigs.

Pool film

We are aware that in the category of pool film – film of common origin used by both BBC and ITN – there was a further degree of failure in the logging sheet because in the event this proved difficult to note. But one point can be made in general. With the single exception of the EBU news exchange, the networks seek different outside sources for film. For instance, they have formal relationships with different international newsfilm agencies – ITN with UPITN, BBC with Visnews. Therefore, the close similarity between channels which has been a continual theme in this chapter has virtually nothing to do with the use of film material from the same source. Only a very small fraction falls into this category and is normally restricted to foreign stories.

Electronic inputs

It must be stressed that the difficulty of noting the distinctions made on the initial logging sheet were, in the area of electronic inputs, even greater than with the use of pool film mentioned above. Our reservations about the significance of the distinctions we initially tried to make in the light of our observation of newsroom practices[20] are compounded by the logging difficulties. But given our caveats about this data, some points can be made as to the distinction between live and videotape inputs. ITN's *First Report*, as we have already noted, has an unusual format which allows time for relatively extended live studio interviews conducted by the presenter. This would explain the generally higher live input noted for this channel as a whole. The format of *First Report* made it easier to spot live interviews than on the other bulletins. Although its format is similar, BBC2's *News Extra* does not have such a high level of live inputs. Phrases such as 'I spoke earlier' helped the noting of the live/videotape distinction for this bulletin. The proportion of interviews is much the same but the late scheduled time of the programme clearly allows for more prerecorded items, often introduced as such.

VTR

News items on videotape originate as studio interviews or transmissions from an Outside Broadcast Unit. Here again, in the event, especially with the studio interviews, this was often difficult to determine from the screen. The political categories include items from both sources. Political coverage requires a minimum of mobility and flexibility. It largely consists in the gathering and presentation of opinion and broadcasters have established facilities (e.g. the BBC's Westminster Studio) to expedite such coverage. Much the same applies, although facilities are fewer, to the home affairs categories, especially local government.

It is interesting to note that this technical category, which essentially relates to calm, studio interviews, is relatively less used in the industrial story category. Politics are of course centralised in ways that industry is not. But this, coupled with, for instance, the similar relative lack of graphics in the industrial category suggests that industrial life is not being accorded a maximum of those technical presentational devices which would contribute to reasoned exposition and argument.

Like many of the other inputs, which represent the use of limited resources of time, effort and money and therefore indicate choices and level of priority, the use of studios outside London and abroad for interviews is particularly instructive and, within our general caveat, easy to note. Here again, the studio is used in the political categories with more frequency than any other. In this technical category we also noted the use of automated studios away from Television Centre in London – Westminster and Broadcasting House (here we relied on direct verbal identification of the studios and recognition of the limited range of backings available in them). We would suggest the studio setting is one which often enhances the authority and personality of the speaker, and its more dominant use in the political category is a noteworthy example of routine practices spinning off possibly unlooked-for consequences – in this case at the level of legitimating hierarchy in society at large.

Apart from its use in the coverage of sporting events the Outside Broadcast is used very selectively. Although, as we noted above, a new level of technology (ENG) is substantially altering this. During the period of our study, as the big 'set piece' of the technology it was not widely used in the context of news. The events most likely to receive this coverage are major speeches, late debates in the House of Commons, press and party conferences, scientific stories (space shots) and unscheduled disasters or acts of terrorism when relatively extended in time. Thus it is very rare but not unknown for an OB unit to cover industrial and economic stories. In our period, a notable example was the coverage of a major vote in the London dockers' strike.

As anticipated, the EBU inputs are found in a restricted range of categories, although not exclusively in foreign news. Even in these categories the input is not large; the vast majority of foreign news is 'home grown'.

The expense of satellite inputs accounts for their limited use. There is one unusual feature of the results which is particularly marked on BBC1. In addition to the expected concentration on sport, foreign news and disasters, satellite was used in the politicians (personal) category. A climax of the Stonehouse story occurred in Australia at the same time as Test coverage. By 'coat-tailing' the sports department's scheduled use of the satellite for cricket, the BBC achieved a number of Stonehouse 'scoops'. This accounts for the use of an expensive input into an unexpected story category.

The radio circuit is more frequently used in the foreign categories, especially by the BBC as would be expected because of the greater number of BBC foreign correspondents. It is the cheapest way of utilising a foreign correspondent. But there are other possible uses within the UK which are also included in the technical category, as we noted in Appendix 2. On occasion, such as bomb blasts in London, telephone calls from reporters on the spot will be taken live into the bulletin in order to achieve an immediacy of coverage not otherwise possible.[21]

Graphic inputs

Used with even greater frequency than film, photographs, maps and graphics provide a major way to break up the *talking head*. However, their use is not determined by a film code. They depend entirely on the journalistic structure of the programme. Their distribution by channel and by category is shown in Tables 4.15, 4.16 and 4.17. The importance of this distribution is that it provides further evidence of the kind of treatment each category of news is likely to receive. As with film and other technical inputs, the use of graphics remains steady from programme to programme, from week to week.

The most common labelling device is a supercaption. It is used for both interviewees and television personnel, hence the distribution between categories reflects the general distribution of persons appearing. A notable difference here is that whereas the BBC uses supercaptions in 15 per cent of all items, ITN uses them in 28 per cent.

Table 4.15 Graphic inputs by category (BBC1, all bulletins). Figures show the percentage of items in each category which contain the given input and are rounded to the nearest 1 per cent.

Input	Category: Parliamentary	N. Ireland	Terrorist	Other Political	Industrial	UK and World	Foreign	Economics	Crime	Home Affairs	Sport	Human Interest	Disasters	Science	Average (all categories)
Caption super	20	14	19	27	17	16	12	14	15	21	8	5	15	20	16
Picture	57	25	32	49	27	32	27	18	48	40	21	31	21	35	33
Graphic	15	3	4	9	7	4	1	35	5	10	33	1	2	—	9
Animated graphic	—	1	—	—	—	1	—	*	1	—	—	—	1	—	*
Map	8	16	12	2	6	13	33	7	8	4	2	5	35	5	11
Wire	—	—	1	1	—	4	6	—	—	—	3	2	4	5	2
RC caption	4	3	1	—	—	9	5	5	—	—	1	—	—	—	2
n =	86	139	69	185	427	168	363	173	114	138	165	114	145	20	Total 2306

* < 0.5 per cent

Table 4.16 Graphic inputs by category (BBC2, all bulletins). Figures show the percentage of items in each category which contain the given input and are rounded to the nearest 1 per cent.

	Category														
Input	Parliamentary	N. Ireland	Terrorist	Other Political	Industrial	UK and World	Foreign	Economics	Crime	Home Affairs	Sport	Human Interest	Disasters	Science	Average (all categories)
Caption super	17	12	17	14	12	17	11	9	4	21	8	5	10	36	14
Picture	56	44	33	69	49	49	45	29	54	73	42	45	21	73	49
Graphic	21	10	—	12	12	8	3	21	13	13	24	—	3	—	10
Animated graphic	—	—	—	1	1	1	2	1	—	1	—	—	—	—	1
Map	6	13	25	—	3	14	31	11	9	4	2	3	38	9	12
Wire	2	—	—	2	—	8	6	2	—	—	6	3	7	18	4
RC caption	4	1	—	—	—	5	3	—	—	—	—	—	—	—	1
n =	54	77	24	100	201	93	212	79	46	69	50	58	61	11	Total 1135

Table 4.17 Graphic inputs by category (ITN, all bulletins). Figures show the percentage of items in each category which contain the given input and are rounded to the nearest 1 per cent.

Input	Category: Parliamentary	N. Ireland	Terrorist	Other Political	Industrial	UK and World	Foreign	Economics	Crime	Home Affairs	Sport	Human Interest	Disasters	Science	Average (all categories)
Caption super	39	27	32	29	36	34	20	29	25	42	16	18	24	18	28
Picture	61	45	32	47	40	50	40	33	55	41	45	26	28	38	41
Graphic	14	1	3	8	8	5	3	34	9	7	38	3	4	24	12
Animated graphic	9	2	1	6	3	—	1	—*	4	1	—	1	—	9	3
Map	5	17	12	9	6	18	41	3	7	3	1	3	30	24	13
Wire	1	1	1	1	—	6	7	1	2	—	8	4	6	3	3
RC caption	—	2	—	2	—	4	1	—*	2	—	2	—	—	—	1
n =	111	161	78	182	404	190	480	167	123	151	172	121	127	34	Total 2501

* < 0.5 per cent.

This difference of style reflects a difference of presentational policy: BBC uses more verbal introductions and acknowledgments while ITN relies more on the visual. The graphic category here applies to television personnel as well as interviewees. Although this is a relatively minor feature of this output, there are indications that the use of captions follows certain rules of presentation. Thus, for example, neither very well known interviewees (such as the Prime Minister) nor *vox pops* are likely to be named in this way. Their status and their function in the story will be signalled in other ways. The supposition that the use of the supercaption is a status indicator is indicated by the relatively few uses noted in the categories of sport, human interest, disasters and crime. Its use is higher towards the 'quality' end of the spectrum. This is perhaps confirmed by the fact that on BBC2, science – a category which in its placement shares many of the characteristics of the above group – here is clearly distinguishable. Super captions were used in 36 per cent of BBC2's science items. They legitimate the authority of science. Because of the general tendency to use more super captions in ITN, such conclusions are more difficult to draw in their case.

There is also a significant between-channel difference in the use of photographs. BBC2 carries most; ITN carries more than BBC1. This relatively heavy use of photographs on BBC2 arises from the untypical formats of news programmes on this channel. A large proportion of pictures are portraits of news personalities, hence the concentration to be noted in the parliamentary and 'other political' categories. Unlike the short BBC1 lunchtime bulletin, the BBC2 *Newsday* bulletin seems to seek photographic illustration and *News Extra* fully illustrates its political coverage. In crime, victims, defendants and locations are treated photographically (we have therefore included here the use of identikit drawings as photographs).

The caption sequence distinction drawn in Appendix 2 was not properly noted. It is essentially a current affairs device and, as such, not much used in the news. But the results reflect the notation of sequences of still pictures of personalities (e.g. to illustrate a political row) which should have been categorised as photographs. Therefore we have abandoned the distinction and incorporated the caption sequence figures in the picture category.

On all three channels graphic inputs (graphs, tables, etc.) are concentrated in the sport and economic story categories. Scores and results of sporting events are nearly always presented in this form.

The heavy use of graphics in the economic category is more significant because it often reflects coverage of a different order from, say, crime or industrial stories. Some professionals regard graphics as evidence of unimpeachable fact – to be seen by the viewers as 'the voice of God'.

It can be inferred from these figures that crime is very rarely considered globally as a social phenomenon despite the availability of regularly published statistics. Similarly, the industrial category tends to have below average graphic inputs, suggesting that a premium is put on 'events' rather than on hard 'facts' of wages, productivity, accidents, strikes and so on although some of these figures were regularly used during our period (wage rate statistics being the most frequent).[22]

The use of animated graphics – which is an attempt to liven up graphics just as graphics are an attempt to enliven *talking heads* – is very limited, having a less obvious pattern of distribution than other graphics, although ITN has significantly more than BBC. In general, taking these figures with the distribution of film between channels, it is possible to suggest that BBC2, because of its specialised formats, uses less film and more graphics than the main channels; and ITN, while using much the same amount of film as BBC1, uses more graphics. Not unexpectedly, maps are used most frequently in the foreign and disaster categories, reflecting their value in providing a quick source of illustration for stories otherwise difficult to treat 'televisually' because of access.

The use of the specialised radio circuit caption described in Appendix 2 follows, obviously, the use of radio circuits. Since the RC is also used as commentary over film or stills, the level of this input is correspondingly lower.

Two main conclusions can be noted from a consideration of these technical and graphic inputs which override the classification problems described above. Professional evaluation of the relative merits of competing news services depends to a large extent on comparisons in these areas. As a contribution to this debate we would suggest that, as against its main rival, ITN attempts a more visual bulletin; it uses more graphic inputs of all types; and although it uses overall less film, more of this is voice-over commentary only. The second conclusion that can be drawn from the whole of this account thus far is that the news on all channels has developed a range of techniques to illustrate all those areas of coverage with which it is concerned and that the relative 'filmability' of one area as against another (one criterion offered as an explanation for the shape of the bulletins by news producers) cannot be used to justify the bulletin profiles.[23]

Interviews

In the three-month period January to March 1975, television news contained interviews with 843 named people. The distribution of these interviewees by bulletin and type as indicated in the initial logging sheet is shown in Table 4.18.[24] The categories 'spokesman' and 'opposition spokesman' are combined for this purpose in view of the lack of grounds for distinguishing between them (see below, Appendix 2, p. 278). Our criterion is that the words spoken (whether in personnel interviews or whether at press conferences, speeches, etc., as in the 'Conference' category) are transmitted as first-person statements. Thus the table provides further detailed evidence of the extent to which the news is dominated by third party (i.e. professional news) reporting and script.

There is one notable difference between channels. A comparison of the main competing bulletins shows that the BBC have a larger proportion of items with interviews in nearly all categories than ITN.

Table 4.18 Interviewee inputs — January–March 1975. No. of items with interviews as a percentage of all items*

Bulletin	Type of interviewee						
	Central figure	Spokes./ opp. spokes.	Vox pop	Witness	Conference	Expert	Total no. of items
BBC Lunchtime	2·6	1·9	0·3	1·3	1·6	—	377
First Report and w/e Lunchtime	6·8	21·4	2·9	1·8	1·6	9·2	546
BBC Early Bulletin	7·0	12·3	0·7	1·1	4·3	1·1	818
ITN Early Bulletin	5·6	8·7	0·8	1·1	2·9	0·3	629
Newsday	1·3	1·5	0·2	0·4	0·7	—	455
Nine O'Clock News	9·4	21·4	1·3	2·8	5·1	1·9	964
News at Ten	8·7	14·7	1·2	1·8	3·9	1·8	1073
News Extra	10·8	26·7	2·5	3·6	4·4	12·5	472

* Includes all weekend bulletins except *News Review*.

But the pattern of the inputs is closely similar. Spokesmen and opposition spokesmen are more frequently interviewed on both main channels than either central figures or witnesses, for example, and *vox pops* are used with even less frequency. Experts are regularly interviewed on only two bulletins – *First Report* and *News Extra* – and not simply because these bulletins include more interviews as a whole but because they both represent a departure from conventional news towards current affairs.

The 843 named interviewees made a total of 1,771 appearances. The distribution is highly skewed in a way which indicates that only a minority of interviewees are in the news for more than one day. 711 (84·3 per cent) of interviewees appeared on only one day and were, of course, drawn from a wide range of the population. Only 132 (15·7 per cent) appeared in bulletins on two or more days and of these, a majority on only two days. The minority of figures who kept reappearing in the news consisted almost exclusively of high-ranking politicians – in order of the number of days on which they appeared: Harold Wilson (21), Margaret Thatcher (17), Tony Benn (14), Michael Foot (10), Edward Heath (10), James Callaghan (9), Denis Healey (8). Also appearing with great frequency were John Stonehouse (12 days on BBC, 2 days on ITN), Robert Booth, the detective in charge of the Whittle kidnap case (11), and Henry Kissinger (10). The most frequently appearing trade unionists during this period were Len Murray and Jack Jones, who were both interviewed on 7 days. On the management side, the most frequently appearing spokesman was Campbell Adamson of the CBI, interviewed on 6 days.

The tendency to interview a restricted range of people can also be seen in the proportion of women interviewees. Only 65 or 7·7 per cent of the total were women and they made 8·4 per cent of appearances. Their distribution throughout the range of categories is uneven, with a relatively higher level of representation in the disaster, politicians (personal), and sports categories. Mrs Thatcher's central place in the Conservative Party leadership story accounted for nearly a third of all appearances by women. The industrial and economic categories, on the other hand, contained very few interviews with women.[25] Thus the coding of news (that it talks to the powerful) reflects society in that women do not hold positions of power.

It was once argued that a lack of balance would be displayed through inequalities in the amount of time or the number of interviews devoted to different points of view, and the news organisations are clearly anxious not to show obvious partiality in this way. The interviewee list provides some data at this level for beginning to examine the somewhat crude notion of balance implied by this charge.

In the industrial category this data has several noteworthy features. The majority of news *vox pops* occurred here but the use of experts was below average; another element, we would argue, in the tendency to seek 'photogenic discord' which characterises industrial coverage. The balance between named interviewees speaking for management (plant level and senior management, employers' associations, owners, CBI) and those speaking for labour (trade union spokesmen, unofficial labour spokesmen, TUC) in the industrial category is not strictly maintained. In the three-month period there were only 33 of the former, but 58 of the latter. There were nearly twice as many official trade union spokesmen as unofficial, and, on the management side, an insignificant number of management spokesmen at the individual factory or plant level.

Another feature of the industrial coverage – and a further indicator of its importance in the news – is that it makes use of interviewees from a wider range of the population than any other category. Thus, politicians are asked to comment on industrial stories with the same frequency as senior management, whilst spokesmen for a variety of interest groups are interviewed with hardly less frequency. The finding that management appear less frequently than labour in the industrial category needs to be qualified although it does confirm a commonly held business view of the coverage.

First the economics (business) category reveals a very different picture and to assess the 'cultural fairness' of access this needs to be taken into account. Here there are 28 spokesmen for management and only 4 for labour, and, in contrast to industrial stories, experts are used with relatively high frequency. Second the distinctions in presentation between industrial and business stories noted above allows for differences of treatment, the effect of which is to give the activities of business (including profits, export orders, bankruptcies) a coherence of treatment which is at the very least not influenced by the search for 'photogenic discord'.

A further difference in the use of management spokesmen is that they are more likely than labour spokesmen to appear in stories marginal to business performance and industrial relations. They figure more frequently in industrial accident stories, for example, and were interviewed on a number of occasions in connection with price increases. The effect of this on the total number of interviewees for management and labour is to put management slightly ahead so that 7·9 per cent of all interviewees were spokesmen for labour and 9·6 per cent spoke for management. It should also be noted that in the industrial area balance, especially in dispute situations, is often sought not between management and workers but between groups of workers with different attitudes to the dispute. Thus the count for labour appearances is more likely to contain extreme differences of opinion (for example the 'Cowley wives' episode during the engine tuners' strike) than is the management count. And concomitant with this is the tendency we think might exist to quote management more often than labour in the body of the news text, a preliminary finding we shall deal with in the next volume. All these aspects of 'balance' have to be seen, we would submit, in the context of the 'facts' versus 'events' we refer to above (Chapter 1).

The choice of interviewees, their selective use in different story categories and the frequency with which interviews are used, once again reveals stability through time and across channels. These patterns create an impression of familiar figures who reappear and provide continuity in an endless sequence of similar but unconnected names and faces. The coherence of this pattern – like the coherence of the bulletin in general – is not to be found at the level of individual appearances or stories but in the association of different types of interviewee with different categories of story and in the use made of interviews in explanation and interpretation.

We have argued in this chapter that the stylisation of the news and similarity between the channels is not to be explained by the pattern of external events as though each channel were reflecting the same image of an external reality. Although some events impose themselves in this way, the outline of the image is prefabricated according to internal, journalistic criteria. The work of the journalist largely consists in filling in this outline. The minor differences between channels which we have noted, however, tend to divert attention from this fundamental manufactured quality of news. The changes made in the news on BBC in March 1976 (simplified titles, a single newsreader and a plainer background) are one example of this. Styles of presentation, especially the varying uses of film and graphics between BBC and ITN contribute to an appreciation of the news and help to account for ITN's reputation of having a less solemn, lighter touch and a more interesting and rounded approach. They do not contribute to an understanding of the news as a cultural, determined product.

5 CONTOURS OF COVERAGE

> No aspect of communication is so impressive as the enormous number of choices and discards which have to be made between the formation of the symbol in the mind of the communicator, and the appearance of a related symbol in the mind of the receiver.
>
> Wilbur Schramm

News selection

The notion of 'the gatekeeper function' is one of extreme importance in considering the televising of news. Even more than in print media the editorial process whittles down, from the vast range of events occurring on any given day in the world, an extremely small number of discrete stories which go to make up the bulletins. In answer to the charge of extreme and idiosyncratic selection the television newsman often responds with what might be called 'the time defence', that is to say the length of the bulletins heavily constrains the number of stories that can be carried and this in turn determines the nature of 'the gate'. We have seen above[1] how the regularities of the bulletin express this production ethic on a day-to-day basis and on what occasions the pattern is disturbed; normally by the process we have called, following Frank, 'swamping'.[2] In the debate about the possibility of extending the news bulletin's time – a demand made in the name of increasing the understandability of the stories by giving more background information – the professionals concerned indicated that if they were indeed given more space they would be inclined to increase the discrete number of stories covered rather than the length of the items within the bulletins.[3]

It then becomes reasonable to suggest that 'the time defence' is less an acknowledgment of the technical restrictions of the form involved but rather a legitimation of an editorial function which effectively conceals its character. What is hereby offered therefore is not a critique of the practices of the television newsrooms in terms of a theoretical model of journalistic requirements but rather a critique which questions the assumptions underlying selection processes. To do this we shall map the contours of television's coverage of industrial life in the United Kingdom during the first 22 weeks of 1975.

A news bulletin is organised as a sequence of items and can vary in terms of its internal complexity. An item may be a simple report.[4] An item may constitute several reports often of only one sentence each. These may be of several different industries and each will be reported.[5] It may constitute a more complex grouping or 'packaging'. For example, an item on the car inudstry may bring together reports on a strike in one car plant, car production figures and foreign imports of cars to the UK.[6]

The basic element is therefore the *report* of which there is generally only one relating to a particular story in a single *bulletin*, but perhaps several reports each relating to different stories within an *item*. A story may run over several days or even weeks and may appear in more than one report. It may be meshed into news items in various ways. The simplest example is the equivalence of one report = one item = one story. In practice this is rare. A story is usually reported more than once. When it is reported, it may take up the whole of an item, or be part of an item. So, for example, the Glasgow dustcart drivers' strike was reported many times, sometimes in items about a number of ongoing Scottish strikes, sometimes in conjunction with the Liverpool dustcart drivers' strike, and sometimes by itself as a whole item. Therefore, identifying and tracing stories involves a good deal of painstaking unravelling because of the variable manner in which news items are constructed. Nevertheless, using transcripts made of the broadcasts it is possible to do this in a systematic way.[7]

We are not importing a notion of the 'news story' as an *a priori* category – it is embedded in journalistic practice.[8] The organising of reporting around certain 'angles' resulting in what we have called the dominant view of an event is not our concern here. The analysis at this point primarily focuses on *which* events are selected for reporting, and how often. To classify these industrial items, we will make use of the Standard Industrial Classification (SIC) (1968). This categorisation is represented in a number of tables throughout the chapter and is routinely used in the Department of Employment

Gazette.[9] SIC operates in various ways but there is one major division by industry (industrial order level); another subdivides this industrial order level into subcategories known as the minimum list headings. For most purposes it has been sufficient to operate at the industrial order level, but we have had occasion to go down to the minimum list headings, notably in the cases of motor vehicles, transport and communication, and public administration. The use of the SIC at once opens up possibilities of using a number of statistical indicators of 'real activity' – which offer publicly available information as to the distribution of events in the industrial world: such as the magnitude of employment, the incidence of industrial disputes, industrial accidents, the agreements on industry-level wage settlements. For some of these purposes we have used the SIC grid with respect to the total five-month period.

By using the SIC, not all industrial stories covered by the bulletins are analysed. A number of industrial stories are reported within bulletin items organised around such overall themes as 'Unemployment' or 'the Social Contract'. The 'packaging' of particular industrial reports in this manner is more common in ITN's *News at Ten* and BBC2's *News Extra*. The industrial story reports have been included in the analysis below while the reporting of the Social Contract, unemployment and the Industry Bill, are assessed as separate stories. Obviously any total assessment of TV coverage of industrial news must necessarily take these latter into account.

The extent of individual stories and the range and distribution of overall coverage between the three news services, industry by industry, is an important component of the basic contours. What follows briefly is the pattern of story coverage as it relates to each industrial grouping (see also Table 5.1).

Agriculture and fishing

In agriculture there were ten news stories, in fishing eight. The only agricultural story to get extensive treatment was the protest by egg producers against imports, sporadically reported throughout February, March, April and May. The remaining nine stories were not extensively covered and, as we shall suggest, this lightness of coverage is reflected in the fact that many of these were exclusives to one channel or another, 4 to BBC1, 2 to BBC2 and 1 to ITN. Of the total of 14 reports on 6 days, 7 were on BBC1 on 5 days, 6 on ITN on 4 days and 1 on BBC2 on 1 day.

Table 5.1 Industrial News Coverage January–May 1975

	BBC1			BBC2			ITN			TOTAL		
	Stor.	Reps.	%	Stor.	Reps.	%	Stor.	Reps.	%	Stor.	Reps.	%
Agriculture	7	15	1·8	4	5	1·2	4	10	1·3	10	30	1·5
Fishing	8	50	5·9	6	23	5·4	5	36	4·6	8	109	5·3
	15	65	7·7	10	28	6·6	9	46	5·9	18	139	6·8
Mining	7	41	4·8	5	23	5·4	5	33	4·2	7	97	4·7
Food, drink and tobacco	–	–	–	2	2	0·5	3	3	0·4	5	5	0·2
Coal and petroleum prod.	6	19	2·2	6	12	2·8	9	22	2·8	13	53	2·6
Chemicals and allied ind.	3	8	0·9	2	3	0·7	6	14	1·8	6	25	1·2
Metal manufacture	7	35	4·1	8	25	5·9	6	31	3·9	9	91	4·4
Engineering	6	31	3·6	3	14	3·3	7	28	3·6	7	73	3·5
Shipbuilding, etc.	5	7	0·8	2	3	0·7	3	6	0·8	5	16	0·8
Motor vehicles	40	213	25·0	24	93	21·8	38	196	24·9	55	502	24·4
Aerospace	3	4	0·5	1	1	0·2	3	7	0·9	3	12	0·6
Textiles	7	13	1·5	5	7	1·6	3	7	0·9	9	27	1·3
Clothing and footwear	–	–	–	1	1	0·2	1	1	0·1	2	2	0·1
Bricks, pottery, glass and cement	1	1	0·1	–	–	–	–	–	–	1	1	0·05
Paper, printing and publishing	5	28	3·3	6	20	4·7	7	36	4·6	8	84	4·1
Other manufacture	–	–	–	1	1	0·2	–	–	–	1	1	0·05
Construction	9	25	2·9	9	18	4·2	7	21	2·7	13	64	3·1
Transport and communications												
Railways	5	47	5·5	3	28	6·6	7	67	8·5	7	142	6·9
Roads	5	14	1·6	3	8	1·9	4	8	1·0	6	30	1·5
Sea	8	30	3·5	6	15	3·5	5	19	2·4	9	64	3·1
Docks	4	34	4·0	4	14	3·3	3	41	5·2	7	89	4·3
Air	6	28	3·3	7	15	3·5	5	23	2·9	9	66	3·2
Postal	3	5	0·6	2	4	0·9	2	8	1·0	3	17	0·8
	31	158	18·5	25	84	19·7	26	166	21.0	41	408	19·8
Gas, electricity and water	2	18	2·1	2	13	3·0	3	13	1·6	3	44	2·1
Distributive trades	3	8	0·9	2	2	0·5	1	4	0·5	3	14	0·7
Education	8	23	2·7	4	5	1·2	6	10	1·3	12	38	1·8
Medical services	7	30	3·5	5	15	3·5	6	31	3·9	9	76	3·7
Public administration	15	89	10·4	10	35	8·2	17	96	12·0	22	220	10·6
Professional, scientific, business services and public admin. —total	32	142	16·6	19	55	12·9	30	137	17·2	43	334	16·1
Miscellaneous services	6	32	4·3	5	22	5·2	6	19	2·2	9	73	3·2
Total	188	848	99·8	138	427	100·1	173	790	99·9	261	2065	99·8

Apart from single reports on the agricultural workers' pay claim and the NUF conference, the main theme of the coverage was the anxieties of farmers on the issues of subsidies, prices and imports.

Although fishing had fewer stories, eight, it was much more frequently reported; a ratio of stories to report of 1 to 14 as opposed to a ratio of 1 to 3 in agriculture. Apart from two stories early in the year involving the loss of one trawler and another going aground, the main theme, as in agriculture, related to the economics of the industry. Five other stories were covered by both ITN and BBC1 and four of them were also covered by BBC2: the lost trawler (January); the financial troubles of Associated Fisheries (February); the French fishermen's blockade of the Channel ports (February); and the UK fishermen's blockade of British ports (March and April). It should be noted that the French fishermen's blockade was more extensively covered on the BBC's channels – by no means an automatic result of having more time than ITN.

ITN carried 4 fishing reports on 2 days, BBC2 carried 5 on 4 days and BBC1 carried 9 on 4 days. In addition the BBC had three exclusive stories: the Aberdeen herring men's demand for a 50-mile fishing limit (February BBC1 and 2); the international conference on fishing conservation held in London (May BBC1 and 2) and the action of fishermen from the Irish Republic in support of the blockade of UK ports (April BBC1). The heaviest coverage within this sector was the UK fishermen's blockade in March and April, which was mentioned in a total of 69 bulletins. This was a novel form of protest and the ample opportunities for aerial photography were fully utilised. The importance attached to this story by all three news services is evidenced by the similarity of coverage: BBC1 provided 28 reports in 13 days; ITN, 28 in 12 days; and BBC2 had 13 in 10 days.

Mining

Of the seven mining industry stories, six of them were given light coverage, never more than two bulletin reports per news story. Three stories on questions of production and a new coal mine in Yorkshire were BBC exclusives. All channels reported the deaths of 28 African miners in a riot, the only foreign mining story. Concern was also reported by all channels over the possible prospect of the 'one hundred pounds a week miner' at home. The story which accounts for the greatest number of references is the miners' pay claim (January and February). This was followed through over five weeks to the final settlement on 28 February. The coverage of this story shows the general similarity between BBC1 and ITN, a characteristic of major stories: BBC1 – 32 bulletins on 15 days, ITN – 27 on 12 days and BBC2 – 19 on 13 days.

Food, drink and tobacco

The coverage of this sector was not extensive. Five unrelated stories, three exclusive to ITN and two to BBC2, received one report apiece; Tate and Lyle profits; increases in bread prices; short time in the tobacco industry; the dangers to infants of dried milk, and the threat of nationalisation of UK interests in the Portuguese wine industry. This shows that exclusivity in running minor stories is not confined to the BBC because of the greater number of bulletins they have available. Here ITN ran three exclusives while BBC1 ignored these industries.

Coal and petroleum products

The year opened with the financial collapse of the Burmah Oil Company, covered extensively by all three channels in a total of 22 reports in 7 days. There was also a report of falling profits at Shell given by BBC1 Early Evening Bulletin on 15 May. However, in April all channels brought us the good news, with correspondents all sending film reports from Alaska, of British Petroleum's oil field development.

The dominant concern in reporting this industrial sector was the development of North Sea oil: the prospect of government control, taxation, oil platform building, job prospects and work hazards, and a new oil strike close to the Shetlands. Covering the North Sea, ITN had 5 exclusive stories; BBC2 had 2; and BBC1 had 1. It also shared one other story with its sister channel. Most of these stories appeared in only one or two bulletins.

Chemicals

News from the chemical industry mainly concerned explosions – both of profits and plants. ITN reported the high profits of Unilever (May), and LaRoche (March), and also carried the story of an explosion of a chemical factory in Kent. Both BBC1 and ITN reported an explosion in a Belgian chemical plant in Antwerp (February) and all bulletins except *News Extra* reported the poisoning of workers at the Cyanamid factory in Gosport on 4 February. The Flixborough accident enquiry was reported by all channels. Amidst the generally light coverage of chemicals Flixborough still dominated.[10]

Metal manufacture

Nine stories are reported in this category but coverage of the British Steel Corporation's manning proposals and their implications for closures and redundancies dominated the news. There were a number of elements to the story but the conflict between Sir Monty Finniston, the chairman of BSC, and Mr Tony Benn, the then Minister of State for Industry, was most widely reported – 24 reports in 11 days on BBC1; 21 on 9 days on ITN; and 13 on 11 days on BBC2. The even coverage between BBC1 and ITN, for what is seen as an important story, is again evident. The remaining lightly covered stories related to closures and layoffs with some attention being given to pay claims. There were 8 of these, and, as happens with less important stories, there were more exclusives, 2 to BBC and 1 to ITN – a foreign story on the Swedish steel industry and the EEC. The only non-steel metals story was a report on *News Extra* on 14 March, of a fire which had been burning for 8 years in a waste tip at the Rio Tinto Zinc plant in Bristol.

Engineering

Seven stories were reported from the engineering industry. One was exclusive to ITN, a progress report on the Fisher-Bendix co-operative. In addition to stories on the engineering workers' pay claim and Ferranti's request for government financial aid, there were two further stories which received extended treatment. One was the sit-in at the Imperial Typewriters factory in Hull following its closure by the multi-national Litton Industries. The other was a strike at the Dunlop engineering plant in Birmingham. Here 700 clerical workers struck on 18 April in pursuit of pay claims. The strike continued into May and led to the layoff of 12,000 workers in the motor industry, principally at BLMC. BBC1 ran 8 reports on Imperial Typewriters on 5 days in January and February, and up-dated on 30 May in the *Nine O'Clock News*; BBC2 ran 3 reports on 3 days; and ITN 6 on 3 days. For the Dunlop strike (April and May)

BBC1 ran 16 reports on 7 days; BBC2 9 reports on 8 days; and ITN 16 on 10 days.

Aerospace

Three stories were covered. A strike at Herne Airport British Aircraft Corporation plant in February was carried by BBC1 and ITN in two bulletins each. The effects of defence cuts on the future production of the Harrier jets and possible redundancies was run in all ITN bulletins on 21 February and one bulletin each on ITN, BBC1 and BBC2 in May. The government nationalisation plans for the aircraft industry were also reported on the main evening news by ITN and BBC1 on 15 January.

Motor vehicle manufacture

This industry was more extensively covered than any other – 55 stories and 502 reports – a ratio of stories to reports of 1 to 9. The coverage dominated industrial news in every month except March. In February, it shared its dominant position with the coverage of the miners' pay claim. Eleven strikes were covered, 23 news stories dealt with various aspects of new orders, sales and new models in the industry, including foreign competitors. Two stories dealt with pay claims (not involving strikes) at particular plants, 5 stories dealt with short time in various firms, and 10 stories with questions relating to government financial aid. Eighteen of the stories were exclusive to BBC and 13 to ITN. But, again the exclusives were not repeatedly referenced and often were included in only one bulletin. Three stories in particular received extensive coverage: the Cowley engine tuners' strike; the Chrysler pay claim strike in Coventry and the Ryder inquiry into BLMC. A comparison between the coverage on the three channels can be seen in Table 5.2.

Table 5.2 Comparison of coverage of three motor vehicle industry stories

Cowley engine tuners' dispute (Jan./Feb.)	ITN 41 reports on 19 days
	BBC1 35 reports on 16 days
	BBC2 18 reports on 9 days
Chrysler plant pay claim strike (May)	ITN 31 reports on 14 days
	BBC1 37 reports on 16 days
	BBC2 24 reports on 15 days
Ryder and BLMC	ITN 42 reports on 19 days
	BBC1 32 reports on 17 days
	BBC2 11 reports on 10 days

The close similarity between news services illustrated in the more prominent stories is reflected in overall coverage. Table 5.1 shows that reporting of the motor vehicle industry constituted 25 per cent of all industrial reporting on BBC1; 24·9 per cent on ITN and 21·8 per cent on BBC2.

Textiles

In essence, news stories here concentrated on foreign imports, layoffs and unemployment, together with requests for government financial aid. None of the stories received particularly extended coverage. BBC covered more than ITN – six exclusives as against one ITN exclusive. One story was covered both by BBC and ITN. This was concerned with government aid to the industry and its import restriction plans. ITN made reference to this on 15 March and it was taken up by all channels in May. BBC2's *News Extra* exclusively reported a fire at a textile factory, treating this as a further aggravation of the unemployment problem. The Hosiery Workers' conference was reported on BBC1 early evening news on the same day. There were nine stories altogether.

Clothing and footwear

Here there were only two stories in one bulletin apiece. Both were concerned with the problem of foreign imports relating to shoes and shirts respectively. The first was reported on BBC2 and the second on ITN.

Bricks, pottery, glass and cement

Here there was just one reference reported on BBC1 *Nine O'Clock News*, on 25 February, the announcement of redundancy plans by Pilkington's glass company.

Paper, printing and publishing

The eight stories reported in this sector all related to newspaper publishing. Two were exclusive to ITN both in April. There was a report on the closed-shop issue discussed at the NUJ conference (*News at Ten* on the 24th) and a report of a pay claim by SOGAT members working on provincial newspapers (early evening news on the 10th). BBC2 exclusively covered an NGA dispute with the

Peterborough Standard over the introduction of new machinery, on 29 May. The remaining four stories were covered on all channels. One involved NATSOPA and the *Mirror* group, reported through 17–20 January (BBC1, 8 reports on 4 days; ITN, 9 reports on 4 days; and BBC2, 5 on 3 days). The other major Fleet Street dispute in January concerned the NGA and the NPA. Over the period 14–22 January BBC1 gave 9 reports to this on 5 days; ITN 16 reports on 8 days; and BBC2 9 reports on 6 days. The unofficial strike of warehousemen at the *Daily Mirror* (March–April) and the ongoing struggles of the *Scottish Daily News* co-operative (March–April–May) were also reported.

Other manufacture

One report on BBC2 (*News Extra*, 3 January) is subsumed here – the announcement that Goodyear and Dunlop were going on to short-time working. This was linked with the recession in the car industry, but unlike the Dunlop story referred to above, which was concerned with the company's vehicle component manufacturing, this story was exclusively about rubber – in SIC terms 'other manufacture'.

Construction

Thirteen stories were reported. Five of them ran on all channels: a report of official figures showing a slump in private house-building (early January), the serious danger of high alumina cement beams in local authority buildings (May) and UCATT members '20 per cent pay deal' on 21 January. The two stories, however, which received heaviest coverage were the axing of the Channel Tunnel project in January (BBC1 carried 4 reports on 2 days; BBC2 5 reports on 3 days; and ITN 7 reports on 3 days) and the continuing controversy surrounding the jailing of two Shrewsbury pickets. On this last it should be noted that the timing and number of references was different as between channels. Only BBC1 followed the story throughout our period: BBC1 ran 8 reports on 5 days (January, February, March and May); BBC2 5 reports on 3 days (January and February); ITN 3 reports on 1 day (January).

BBC's coverage in this sector was generally more extensive than ITN's. BBC1 reported two accidents, one involving the death of a worker on a Portsmouth motorway site (28 April, and another a fatality in Liverpool (8 May). BBC1 and 2 reported the complaint of

British construction workers in Canada that they had been beaten up by Canadian construction workers and forced to leave (22 February). BBC2's *News Extra* had three exclusives: an early report on the alumina cement scandal in January, a statement on unemployment in the industry by David Basnett, General Secretary of the TGWU on the same day, 7 January, and the Skelmersdale council building site on which sabotage by workers was alleged to have taken place (21 February). ITN's one exclusive story reported on a race to build a brick house in record time (23 January). ITN carried an early report (11 January) on the strike-bound Montreal Olympic site, devoting further reports to it on May, when BBC1 also referred to it (21 May).

Gas, electricity and water

ITN ran a brief item on nuclear health hazards at de-radiation stations (30 January, *News at Ten*). There was a more extended reporting of electricity price rises on all channels in February and March. The only heavily covered story, however, related to the power workers' pay claim. BBC1 was the only channel to raise the matter as early as January. All three channels took it up in March, April and May. BBC1 ran 16 reports on 9 days; BBC2 11 reports on 9 days; and ITN 9 reports on 6 days.

Transport and communication

In this sector, as Table 5.1 shows, there were 41 stories and 408 reports. We have found it necessary to make use of the various minimum list headings which the detailed Standard Industrial Classification subsumes. Seven rail stories gave rise to the largest number of reports within the transport category – 142. Apart from a reference to a rail strike in Japan on ITN, the stories all concerned British Rail's financial problems and the ongoing wage negotiations. We can see from Table 5.3 that three stories received relatively heavy coverage: the signalmen's strike over pay in February and March, the NUR/ASLEF pay claim from February to May and the work to rule of British Rail maintenance workers in support of a pay claim during March and April.

The similarity of coverage between ITN and BBC1 is very marked in the first two of the above stories, but in the case of the maintenance men's work to rule, ITN's coverage is more extensive.

Table 5.3 Comparison of coverage of three rail industry stories

Signalmen's strike	ITN 20 reports on 11 days
	BBC1 18 reports on 11 days
	BBC2 8 reports on 8 days
NUR/ASLEF pay claims	ITN 23 reports on 12 days
	BBC1 20 reports on 12 days
	BBC2 16 reports on 14 days
British Rail maintenance men's work to rule	ITN 14 reports on 7 days
	BBC1 6 reports on 6 days
	BBC2 4 reports on 4 days

Road transport. This received less coverage. The main story was the protest and sympathy strike following the death of a London bus conductor who had been assaulted whilst on duty in January. BBC1 carried 7 reports on 4 days; BBC2, 5 reports on 5 days; ITN, 4 reports on 3 days. Of the remaining five stories, one related to a strike response to football hooliganism on London Transport in May. This was not covered by ITN as it was off the air as a result of the ACTT dispute with the commercial television companies. The only ITN exclusive was a report on the financial troubles of the National Bus Company in *News at Ten*, 15 January. The story of the death by poisoning of a lorry driver working at a waste tip in Essex received moderate coverage. BBC1 reported the redundancies pending in the National Freight Company in two bulletins on 7 May. There was one foreign story – a report of the petrol delivery drivers' strike in the Irish Republic, on BBC1 and ITN in April.

Sea transport. News from the sea began with reports on the loss of the MV *Lovat*, a prominent story covered in 6 bulletins on BBC1, 4 on BBC2, and 6 bulletins on ITN during 25, 26 and 27 January. Another ship lost was the *Compass Rose*, an oil survey vessel, reported by ITN in 3 bulletins in April and May and on BBC1 in 3 bulletins on 16 April. Two other accidents in February were reported; a collision of foreign ships in the channel (BBC2's *Newsday* bulletin, 4 February) and the fatal fire on board the *Pegas* off Guernsey (Lunchtime News, BBC1, 27 February). One story exclusive to BBC1, on 8 February, concerned allegations of theft by the crew of a cargo ship. All channels reported the *QE2* luxury cruise in January and again in April; the seamen's pay claim also in April; and the strike in March against proposed cuts in Sealink ferry services. Whilst ITN was off the air during the lockout of ACTT members in the commercial TV companies, BBC1 and 2 reported a seamen's strike arising from the introduction of a new channel ferry (BBC1, 7 bulletins and BBC2, 4 bulletins on 27, 28 and 29 May).

Port and inland water. The 'container dispute' was clearly the most prominent story, running in 72 bulletins during February, March and April (BBC1, 27 reports on 13 days; BBC2, 10 reports on 9 days); ITN (giving it closer attention) 35 reports on 17 days. Three other disputes were reported: ITN covered stoppages in Hull and Manchester docks in April, and BBC1 and 2 reported a stoppage in Southampton docks in May. Exclusive coverage was given to two further stories by the BBC: a report on the profits of the nationalised ports (May, BBC1 and 2), and a fatal accident in Port Talbot docks in January (BBC2). The pay deal for London dockers, agreed in April, was reported by BBC1 and ITN in 2 bulletins each on the 29th.

Air transport. Only in March did air passenger transport fail to make significant news, with BBC2's *News Extra* alone reporting that British Airways planned to cut its staff by some 1,300. In the other four months, four strikes of airport ground staff received attention. Two were covered by all three news services – a strike at Manchester airport over pay in April (BBC1, 5 reports; ITN, 3 reports; and BBC2, 1 report), and a strike by APEX members over the organisation of the introduction of the Glasgow–London 'shuttle' service in February (BBC1, 8 reports; ITN, 14 reports; BBC2, 6). The opening of this service in January was also covered by the three channels. The attempts of Laker Airway's 'walk on–walk off' flight service to get off the ground were reported exclusively by the early evening bulletin on ITN (7 February). ITN also reported the British Airways' engineering maintenance staff pay claim at Heathrow airport, in three bulletins on 30 and 31 January. The three other stories were shared by BBC1 and 2: the criticisms from BALPA over the handling of the hijack at Heathrow (*News Extra* and *Nine O'Clock News*, 8 January), and two strikes, one of ground staff at Manchester airport in February (BBC1, 2 bulletins; BBC2, 1 bulletin) and the strike of maintenance engineers at Heathrow airport in May (BBC1, 8 bulletins; BBC2, 4 bulletins on 29, 30 and 31 May). (This latter story appeared during the ITN off-air period.) Other airport services stories concerned local authority staffs and are dealt with below.

Postal services. The Post Office Workers' Union appeared firstly during March, in reports of the Department of Employment's criticism of certain clauses in the union's new pay award (BBC1, 2 reports; ITN, 3; and BBC2, 3), and second during the union's conference in May where an important decision on mechanisation and manning levels was taken (total of 5 bulletins, 16 May). ITN carried two further reports concerning the conference discussions of the Social Contract on 19 May. The announcement of increased postal charges was reported exclusively by BBC1 *Nine O'Clock News* on 21 February.

Distributive trades

A strike by Co-operative Society workers was exclusively reported on *Nine O'Clock News* on 9 March. Two other issues which focused on the attitudes of USDAW members to the Social Contract were covered on all channels. These were a speech by Lord Allen, the union's General Secretary, reported by *News Extra* and the early evening bulletins of ITN and BBC1 on 27 February; and the union's conference in April covered in 5 bulletins on BBC1, 4 on ITN and once by the BBC2's *News Extra*.

Education

Reflecting the general interest in union conferences displayed by television news in 1975 (not least because of the importance attached to monitoring the performance of the Social Contract), the conference of the NUT at the end of May was the biggest education story covered. It was reported on 6 BBC1 bulletins and 2 bulletins on BBC2.

Other news items received light coverage, rarely being included in more than one bulletin on any channel. There were four strikes reported; the Buckinghamshire teachers' one-day strike and demonstration against education cuts appeared exclusively on ITN's *News at Ten* on 14 February; a return to work after a two-year strike by teachers in Redcar and the Scottish teachers' strike both covered by ITN and BBC1 in January; and the strike of university teaching staff in May covered by all three channels.

Another demonstration against education cuts, this by teachers in Richmond, was reported exclusively on ITN *News at Ten* on 22 February, whilst BBC1 and BBC2 together reported the teachers' pay claim on 30 April and the NUS grants campaign on 23 May. ITN and BBC1 both covered the elections to the NUS executive (8

April), the acceptance by Scottish teachers of a £6 pay offer (12 May) and reported on the planned government reductions in teacher training programmes.

Medical

During January, February, March and April the hospital consultants' work-to-rule in their dispute over their NHS contracts and subsequently over paybeds, was followed on all channels. BBC1, 17 reports on 11 days: ITN, 17 reports on 13 days; and BBC2, 8 reports on 7 days. This was the dominant story. BBC's *News Extra* (19 March) informed us of a USA doctors' strike, cross-referencing our own consultants' work-to-rule.

The pay negotiations of the junior hospital doctors and GPs were reported on all three channels in January, with ITN including an update in February. Other pay deals within the NHS were also reported: the radiographers' on all three channels and on ITN a mention of the dentists' pay agreement, both in January. All channels reported a 30 per cent pay increase for such doctors as agreed to it in April.

On 21 February BBC1 *Nine O'Clock News* reported that GPs could now prescribe the contraceptive pill on the NHS; all BBC1 bulletins on 23 May told of the protest made by some doctors to the BMA against amendments to the Abortion Act. All ITN bulletins and BBC1's *Nine O'Clock News* on 22 May reported the General Medical Council's ruling on foreign doctors' qualifications to practise in the UK. Other hospital services stories, concerning non-medical employees, are dealt with below.

Public administration

Exactly half of the large number of stories in this sector relate to industrial stoppages. The 11 disputes included 2 with particularly heavy coverage. The Glasgow dustcart drivers' strike January to April had 36 reports on 19 days on BBC1; 43 reports on 22 days on ITN; and 17 reports on 13 days on BBC2. And the Scottish ambulance controllers' strike had 11 reports on 5 days on BBC1; 9 reports on 5 days on ITN; and 5 reports on 4 days on BBC2, through the period 12–19 January.

The strike of Glasgow Corporation electricians, which affected Glasgow airport amongst other places, during the same period as the dustcart drivers' dispute, received much lighter coverage. BBCl carried 6 reports on 5 days; ITN, 3 reports on 3 days; and BBC2, 1 report only. The three news services also reported during March the successful strike of dustcart drivers in Liverpool, ITN giving coverage in 8 bulletins and BBCl and 2 making 3 reports each. A strike by bin-men at Southwark was reported only on BBCl's early evening news on 10 May. The other BBC exclusive was a strike by NUPE members at Christie Hospital, Manchester, given on the *Nine O'Clock News* on 6 March. In February there was a NUPE strike at Morriston Hospital, Swansea, which BBCl reported 3 times on 3 days; ITN, 8 times on 4 days; and 1 report on BBC2.

ITN had 2 exclusive dispute stories. *News at Ten* on 3 February told of a strike by Avon Council workers in protest against computer miscalculation of their wages. On the early evening bulletin of 13 May, the national firemen's pay-claim dispute was reported.

In March there was an unofficial strike of civil servants at Westminster and Whitehall and this received even coverage as between BBCl and ITN in a total of 14 bulletins. ITN and BBCl both covered the unofficial strike of prison officers during April.

The Civil Service pay deal was noted on the main evening bulletins of both BBCl and ITN on 14 April. Three further stories received moderate coverage in May: BBCl ran 4 reports on 2 days; ITN ran 4 reports on 2 days; and there was one mention on BBC2 of the controversy between NALGO and the Secretary of State for the Environment over proposed cuts in government expenditure on local authorities. BBCl and BBC2 ran 3 reports each and ITN 2 reports on the armed forces' pay rise. The NUPE conference, where, as with other union conferences reported, the important news angle was the Social Contract, was covered by BBCl in 6 reports on 2 days; ITN ran 4 reports on 2 days; and BBC2 ran 3 reports on 3 days.

Of the remaining seven stories, ITN exclusively reported the MPs' pay rise on *First Report* on 7 March, a NUPE study of local authority child minders on *First Report* on 10 March and the same union's action over the private patients issue in Oxfordshire and Berkshire on *First Report* and *News at Ten* on 14 April. On 20 May *News at Ten* reported the acceptance of wage cuts by council workers in Bournemouth. BBCl's exclusive coverage was a report on the prospect of redundancies facing civilian workers as a result of defence cuts (19 March, *Nine O'Clock News*) and the police pay claim noted on three bulletins at the end of May. BBC2's *News Extra* alone informed us of a Civil Service Report on Jobs and Conditions (19 February).

Miscellaneous services

The actual coverage in this sector is not as diverse as the catch-all title might suggest. There were five stories in sport and recreation and four in entertainment. These groupings are minimum list headings within the category.

The biggest story by far in sport and recreation was the stable lads' strike in April/May at Newmarket: BBC1, 11 reports on 6 days; ITN, 9 reports on 4 days; and BBC2, 6 reports on 5 days. As with the fishermen's blockade, this dispute was in some respects a novel form of action. Strikes in sport are rare and picketing of race courses hitherto unheard of. Other stories in this group received very light coverage. The lifting of the women professional tennis players' boycott threat to Wimbledon on a pay question was noted on all three channels on 26 February in one bulletin each. The remaining three stories were: the Racing Board's Financial Report (ITN, 29 January); the Aintree owners/Jockey Club/Betting Levy Board dispute which put the Grand National at risk (BBC1, 4 March); the financial difficulties of the National Trust which threatened employees with redundancy (BBC2, 5 April).

In the entertainment category there was one dispute which both made news and cancelled it. This was the ACTT members' dispute in May with the commercial television companies over pay. ITN in three bulletins reported the impending action which left only BBC

Table 5.4 Areas of major industrial coverage expressed as a percentage of total (industry specific) coverage, by channel (January–May 1975)

Industry category	BBC1 % reports	BBC2 % reports	ITN % reports	Total % reports
Motor vehicles	25·0	21·9	25·0	24·4
Transport and communication	18·5	19·7	21·0	19·8
Public Administration	10·4	8·2	12·0	10·6
Total	53·9	49·8	58·0	54·8
Engineering	3·6	3·3	3·6	3·5
	n = 848	n = 427	n = 790	n = 2065

on the air to report the ensuing strike and lock-out, which they did – BBC1, 15 times on 7 days, and BBC2, 12 times on 8 days.

Finally all channels reported the increase in the TV licence fee on 29 January; the main evening news on ITN and BBC1 informed us of projected IBA cutbacks in expenditure. ITN, with understandable loyalty, reported ITCA's evidence to the Annan Commission on the Future of Broadcasting in its two evening bulletins on April.

Table 5.4 draws together a number of salient findings on the distribution of coverage. Taking the three SIC categories that have the largest amount of news coverage, motor vehicles, transport and communication, and public administration, we see that 54·8 per

Table 5.5 Great Britain – estimated number of employees in employment, June 1975

Industry (Standard Industrial Classification 1968)		% of total employed population
Agriculture, forestry and fishing	388·0 (thousands)	1·8
Coal mining	303·8	1·4
Other mining and quarrying	47·2	0·2
Food, drink and tobacco	714·0	3·2
Coal and petroleum products	39·8	0·2
Chemicals and allied industries	425·3	1·9
Metal manufacture	500·3	2·3
Engineering	1871·5	8·5
Shipbuilding and marine engineering	176·6	0·8
Motor vehicles	456·7	2·1
Aerospace	205·2	0·9
All other vehicles	88·6	0·4
Metal goods not elsewhere specified	542·3	2·4
Textiles	503·8	2·3
Leather, leather goods and fur	41·1	0·2
Clothing and footwear	389·4	1·8
Bricks, pottery, glass, cement, etc.	277·8	1·3
Timber, furniture, etc.	261·5	1·2
Paper, printing and publishing	561·9	2·5
Other manufacturing industries	322·3	1·5
Construction	1241·9	5·6
Gas, electricity and water	345·9	1·6
Transport and communication	1499·2	6·8
Distributive	2641·8	11·9
Professional and scientific, finance and business services	4548·0	20·5
Miscellaneous	2119·5	7·6
Public administration	1623·8	7·3
Total, all industries and services	22137·2	100

Source: *Dept of Employment Gazette*, January 1976.

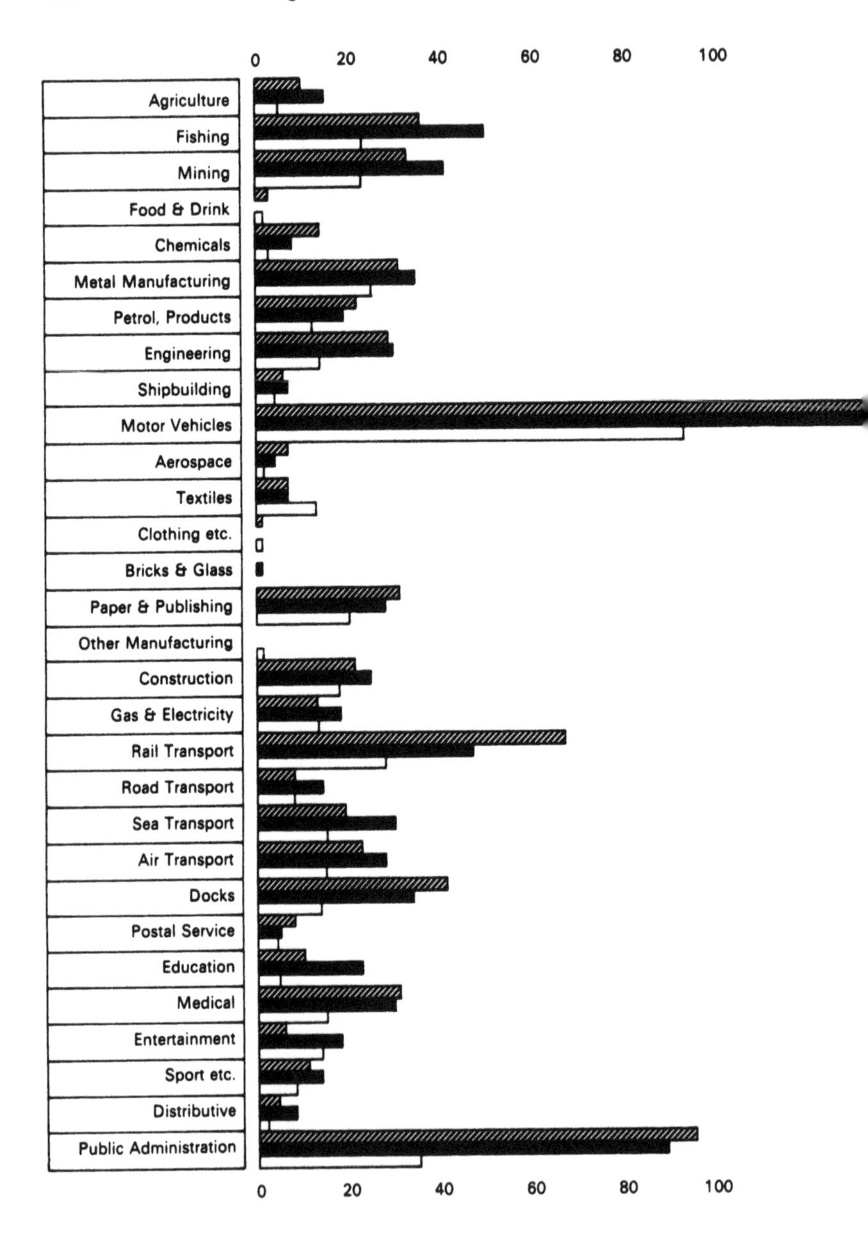

Figure 5.1 Industrial coverage, January to May 1975

Showing distribution of news reports by channel and Industry Sector (D. of Employment S.I.C. 1968)

▸196
▸214

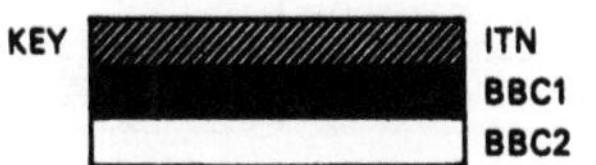

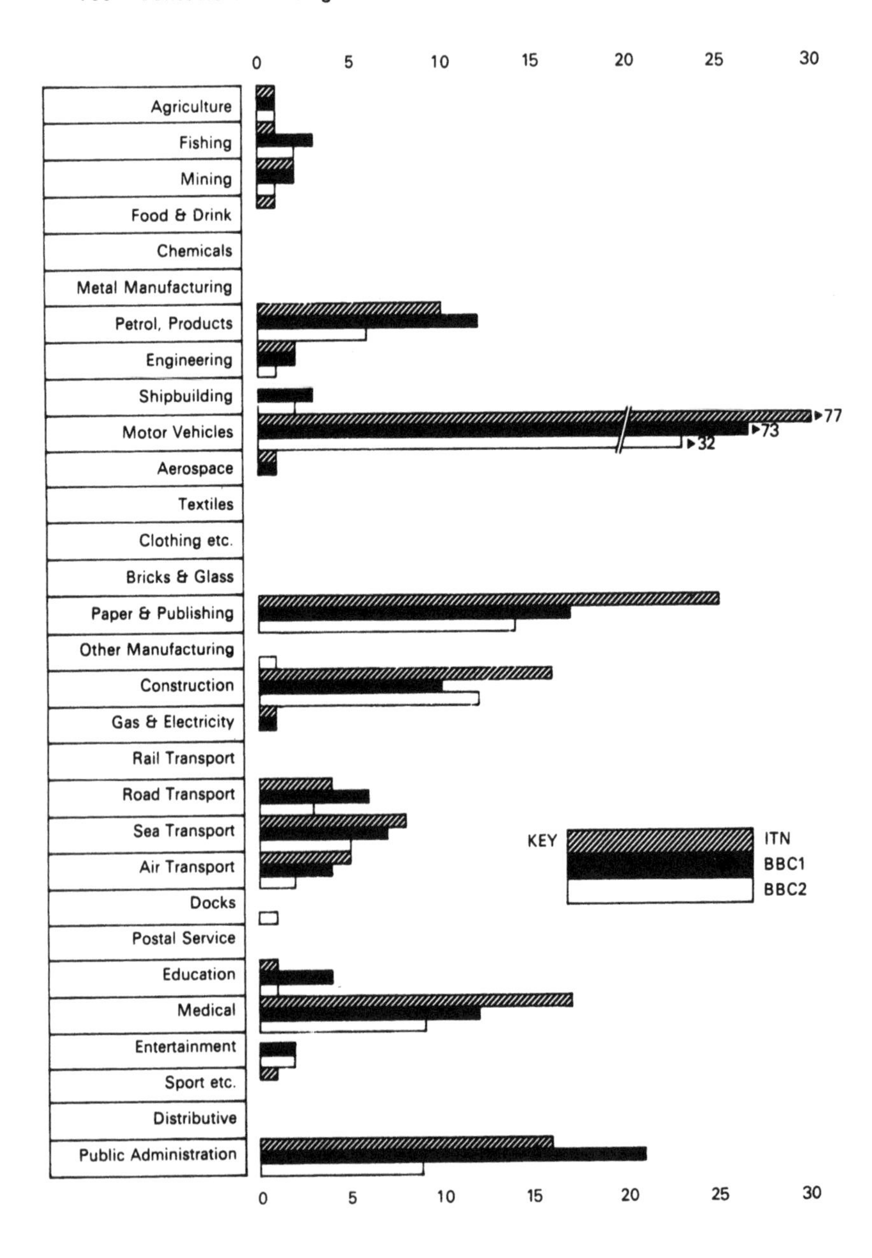

Figure 5.2 Industrial coverage, January

Showing distribution of news reports by channel and Industry Sector (D. of Employment S.I.C. 1968)

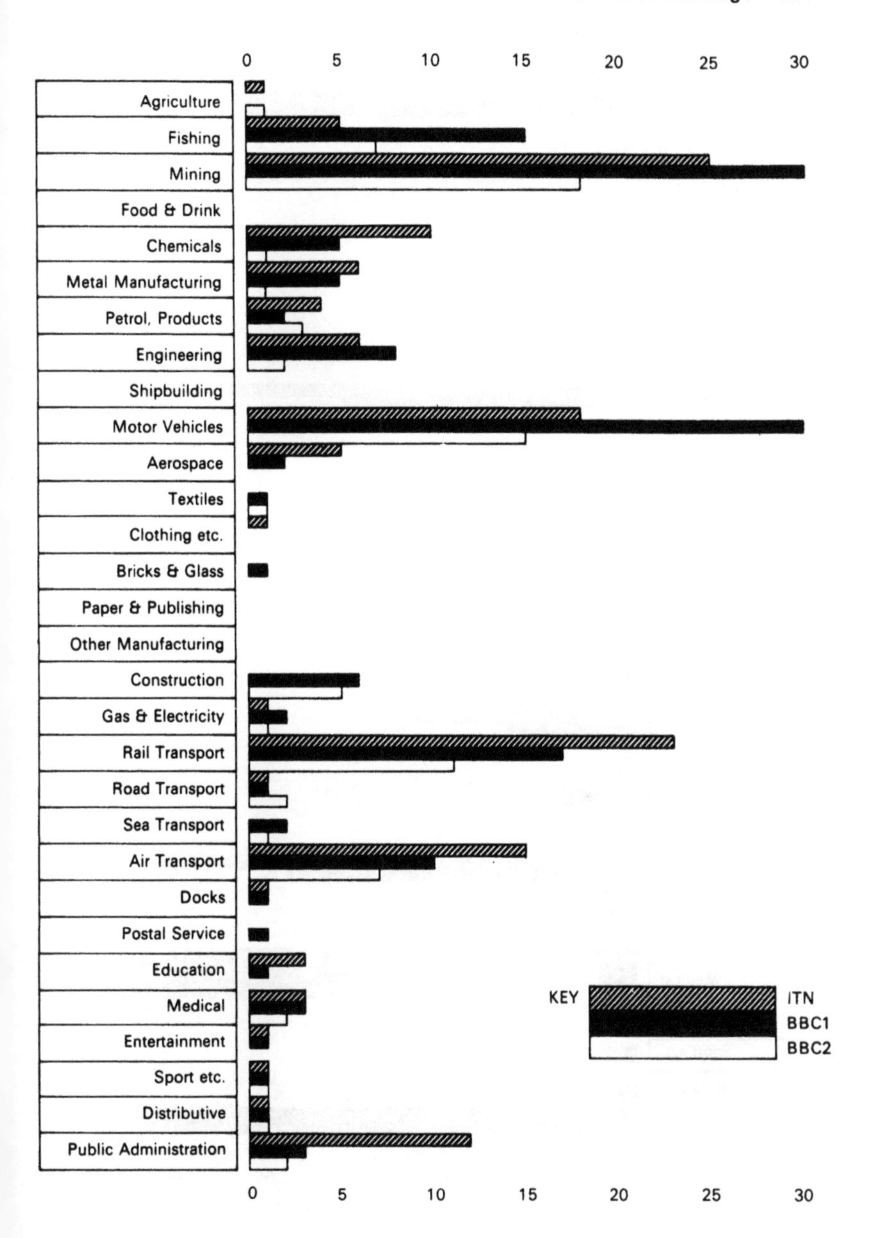

Figure 5.3 Industrial coverage, February

Showing distribution of news reports by channel and Industry Sector (D. of Employment S.I.C. 1968)

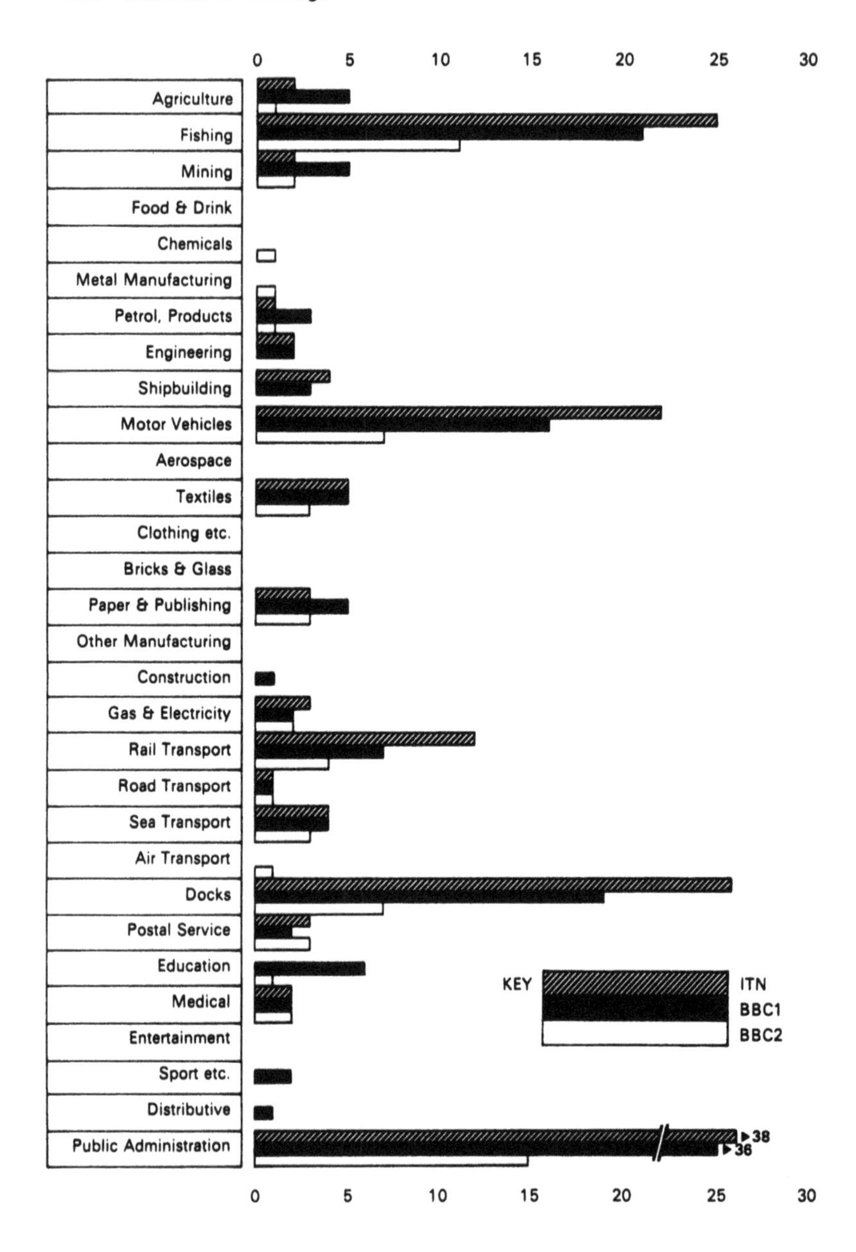

Figure 5.4 Industrial coverage, March

Showing distribution of news reports by channel and Industry Sector (D. of Employment S.I.C. 1968)

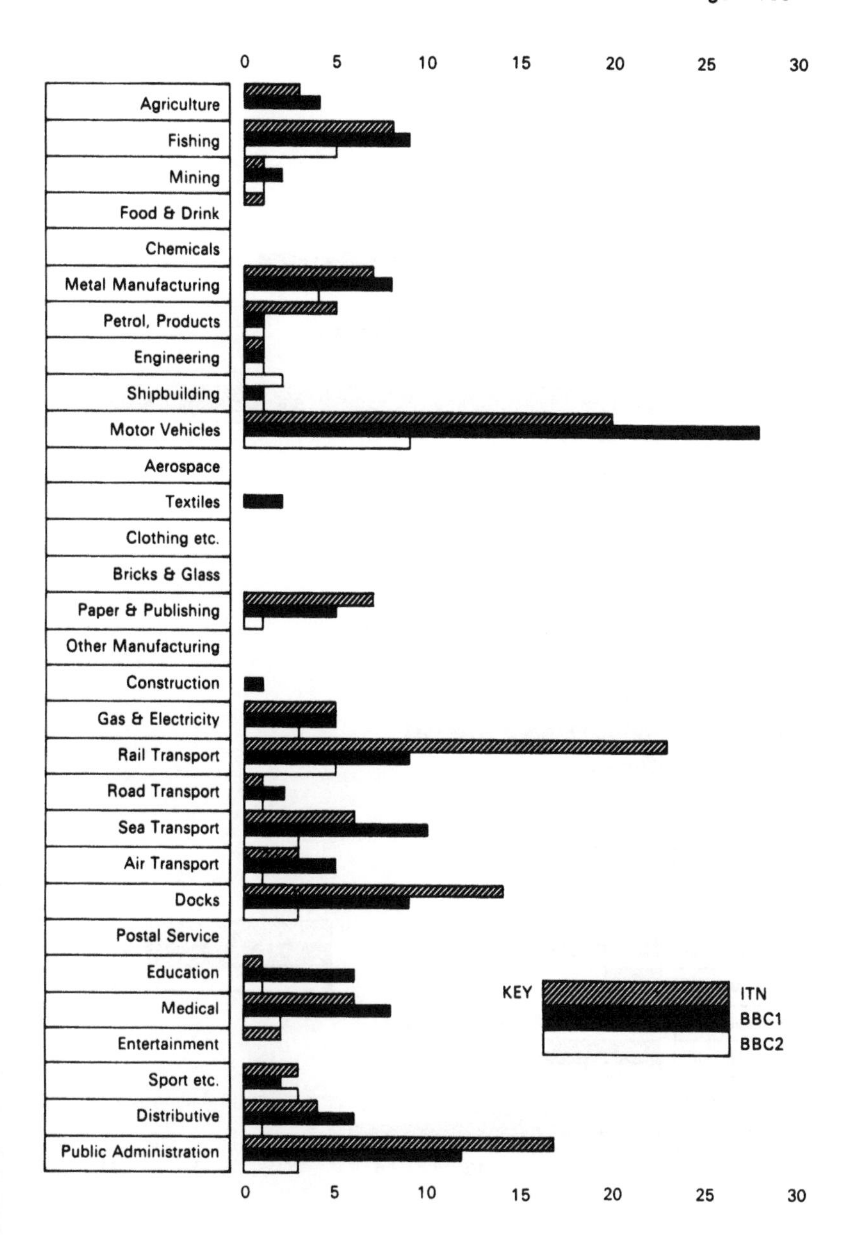

Figure 5.5 Industrial coverage, April

Showing distribution of news reports by channel and Industry Sector (D. of Employment S.I.C. 1968)

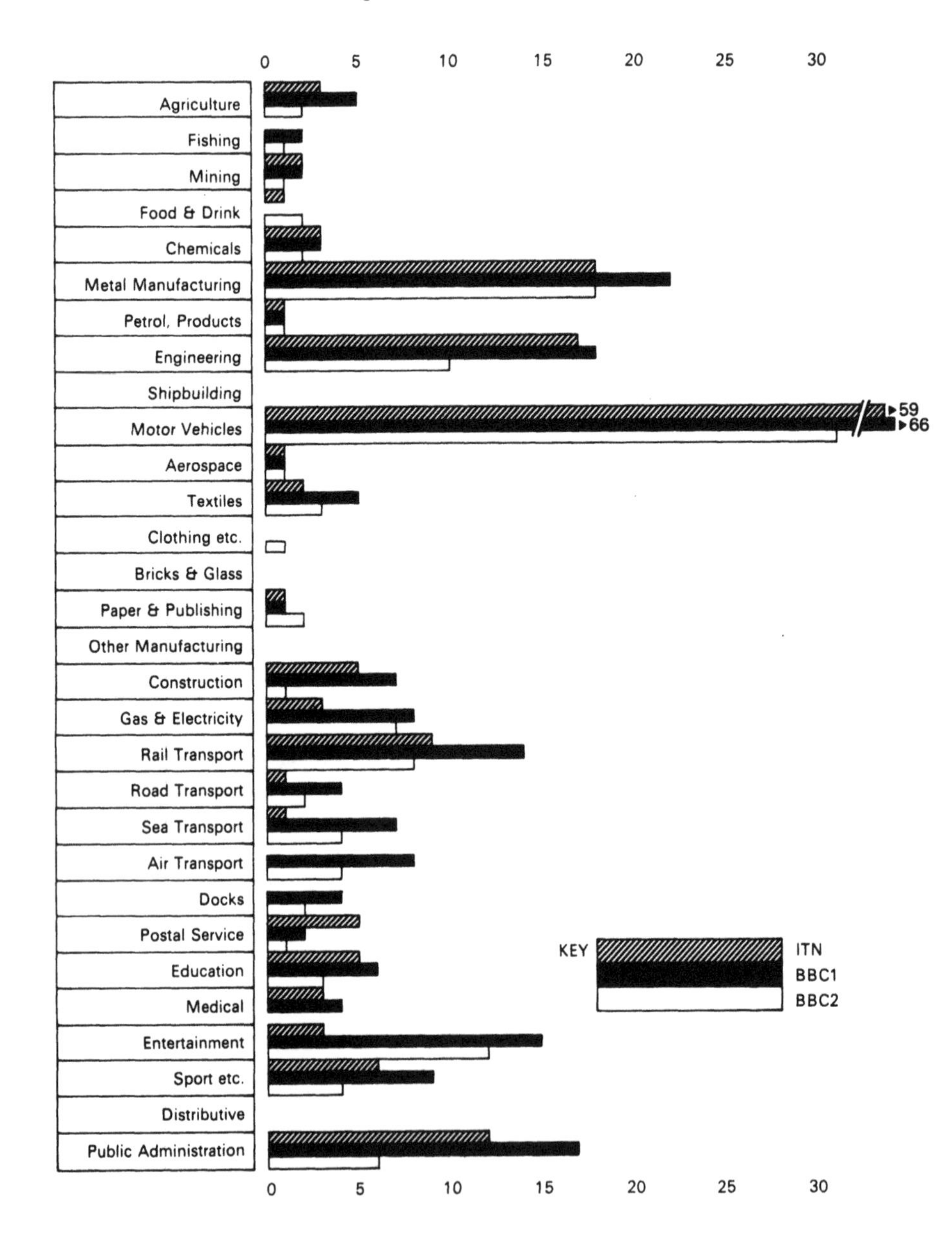

Figure 5.6 Industrial coverage, May

Showing distribution of news reports by channel and Industry Sector (D. of Employment S.I.C. 1968)

cent of all reports are concentrated there. This large proportion is reflected by all three channels, 53·9 per cent on BBC1, 58·0 per cent on ITN, 48·8 per cent on BBC2.

For comparative purposes, we have indicated in Table 5.4 the amount of coverage given to the engineering industry. We have chosen this as it is the category in the manufacturing sector with the greatest number of employees (Table 5.5). The amount of coverage measured by reports is uniformly low in all three channels. This immediately demonstrates that there is no direct relation between the size of an industry as measured by employment magnitudes and the amount of coverage given to it in television reporting. This can be further shown by reference to Tables 5.5 and 5.4. In the case of engineering, we have an industry involving 8·5 per cent of the employed population which makes only 3·5 per cent of industrial news coverage. If we now compare this with motor vehicles we observe that 2·1 per cent of the employed population are represented by 24·4 per cent of industrial news reporting. In the same way, the point applies to transport and communications, where 6·8 per cent of the employed received 19·8 per cent of industrial news reporting.

The reverse side of the case is further illustrated if we look at the overwhelmingly un-newsworthy distributive trades category. Although representing 11·9 per cent of all workers (over 2½ million people) only 0·7 per cent of total industrial news is related to them.

Figure 5.1 shows the extent to which each sector of industry appeared in the news bulletins during the analysis period, January to May 1975. The industries are grouped in accordance with the SIC but only those areas which are covered by news reports are shown. Thus two broad groupings received no coverage – the timber industry, including the manufacture of furniture, etc., and 'metal goods not elsewhere specified'.

Considering those areas which received significant coverage (i.e. average of approximately 10 or more reports per channel), we find at least broad general similarities between ITN and BBC1, in many cases amounting to very close agreement over the industrial areas from which news is drawn. With the single exception of the textile industry, the proportional distribution of BBC2's coverage, despite the untypical structure of its main bulletin, closely follows that of ITN and BBC1. BBC2 has no reports in one sector – bricks, pottery, glass and cement.

BBC1 has no reports in three sectors – bricks, pottery, glass and cement, clothing and footwear, and other manufacture.

ITN has no reports in two sectors, – bricks, pottery, glass and cement, and other manufacture.

Both the channel absences and exclusives are found within the areas of light coverage. What is covered extensively tends to be covered by all.

The major areas of industrial news in the period were transport and communications, public administration (being mainly concerned with hospital staffs and other local authority service workers) and the car industry.

Table 5.6 UK industrial stoppages, strikes, lock-outs—TV coverage January–May 1975

SIC	BBC1			BBC2			ITN			Total		
	Stor.	Reps.	%	Stor.	Reps.	%	Stor.	Reps.	%	Stor.	Reps.	%
Metal manufacture	1	2	0·6	1	1	0·6	1	3	0·96	1	6	0·7
Engineering	2	18	5·3	1	9	5·3	1	16	5·1	2	43	5·3
Motor vehicles	8	95	28·0	5	47	27·8	8	83	26·6	11	225	28·0
Aerospace	1	2	0·6	–	–	–	1	2	0·6	1	4	0·5
	9	97	28·6	5	47	27·8	9	85	27·2	12	229	28·4
Shipbuilding and marine engineering	1	1	0·3	–	–	–	–	–	–	1	1	0·1
Textiles	3	5	1·5	1	2	1·2	–	–	–	3	7	0·9
Paper, printing and publishing	3	22	6·5	4	16	9·5	3	29	9·7	4	67	8·3
Transport and communications												
Rail	1	18	5·3	1	8	4·7	1	20	6·4	1	46	5·7
Road	2	8	2·4	2	7	4·1	1	4	1·3	2	19	2·4
Sea	2	11	3·2	2	7	4·1	1	4	1·3	2	22	2·7
Docks	2	28	8·3	2	11	6·5	3	39	12·5	4	78	9·7
Air	4	23	6·8	4	12	7·1	2	17	5·4	4	52	6·5
			26·0			26·6			26.8			27·0
Distribution	1	1	0·3	–	–	–	–	–	–	1	1	0·1
Public administration and education	3	9	2·7	1	2	1·2	3	5	1·6	4	16	2·0
Local/national government	9	70	20·6	6	29	17·2	9	79	25·2	11	163	20·2
	11	88	23·3	11	45	18·4	8	84	26·8	13	217	22·2
Miscellaneous	2	26	7·7	2	18	10·7	2	12	3·8	2	55	6·8
Total	45	339	339	32	169	169	36	313	313	54	805	805

The distribution of the coverage month by month is shown in Figures 5.2–5.6. The channels show greater variety in the overall amounts and proportions within different industrial areas on a monthly basis. However, the same sectors of industry predominate throughout the period.

Industrial disputes

There is no consistent relationship between the stoppages recorded during the first five months of 1975 (see Table 5.8) and those reported by television news (see Tables 5.6 and 5.7). There are no news stories of the stoppages occurring in seven of the industrial sectors: mining; the manufacture of food, drink and tobacco; coal and petroleum products; chemicals; other manufacturing industries; gas, electricity and water; and the construction industry.[11] However, there were stoppages in these sectors according to the Department of Employment disputes statistics and they accounted for a total of 17 per cent of all the recorded working days lost and 37 per cent of all stoppages in the period.

Table 5.7 Major areas of industrial dispute coverage on TV, expressed as a percentage of total *dispute* coverage (January–May 1975)

Industry category	BBC1 % reports	BBC2 % reports	ITN % reports	Total % reports
Motor vehicles	28·0	27·8	26·6	28·0
Transport	26·0	26·6	26·8	27·0
Public administration	23·3	18·4	26·8	22·2
Total	77·3	72·8	80·2	77·2
Engineering	5·3	5·3	5·1	5·3
n =	339	169	313	805

Further, when disputes are reported, there is no direct relation between the amount of coverage they receive and their severity. For example, whilst the significance of the labour relations problems in the car industry would seem to be reflected in the proportion of coverage devoted to it (reports in 225 bulletins, 28 per cent of all strike reporting), shipbuilding disputes are represented by a single report in only one bulletin on BBC1. This industry recorded 38 major stoppages and 6·7 per cent of the total working days lost. The engineering industry recorded 24·9 per cent of the total days lost in 260 stoppages (of a total of 1,086 stoppages for all industries). Yet engineering stoppages disputes were covered in only two news stories spread over 43 bulletins on all channels. This constitutes only 5·3 per cent of total dispute reporting.

Working days lost due to disputes is not the only possible indicator of the severity of a strike. However, attention to other indices or strike activity – the numbers of workers involved, and the incidence rates of days lost and number of stoppages per worker – also reveals the non-representative nature of reporting in this area of industrial news.

Table 5.8 Number of stoppages (January–May 1975)

SIC	Stoppages beginning in period	Stoppages in progress		Stoppages per worker*	Working days lost per worker†
		Workers involved	Working days lost		
Agriculture, forestry, fishing	—	—	—	—	—
Coal mining	95	12900	25000	31·3	82·3
All other mining and quarrying	1	100	1000	2·1	21·2
Food, drink and tobacco	40	7200	46000	5·6	64·4
Coal and petroleum products	1	600	7000	2·5	175·8
Chemicals and allied industries	34	16300	73000	8	171·6
Metal manufacture	65	24700	102000	13	203·8
Engineering	260	84600	733000	13·9	391·6
Shipbuilding and marine engineering	38	20400	202000	21·5	1143·8
Motor vehicles	72	88900	522000	15·76	1142·9
Aerospace equipment	21	8800	65000	10·23	316·7
All other vehicles	12	9200	132000	13·54	1489·8
Metal goods not elsewhere specified	60	12100	101000	11·1	186·2
Textiles	25	8200	44000	4·9	87·3
Clothing and footwear	17	3700	23000	4·36	59·1
Bricks, pottery, glass, cement, etc.	20	3700	16000	7·19	57·6
Timber, furniture, etc.	13	2300	15000	4·97	57·4
Paper, printing and publishing	19	5900	36000	3·4	64·1
All other manufacturing industries	25	9900	99000	7·7	307·2
Construction	88	12000	92000	7·1	74·1
Gas, electricity and water	9	3600	8000	2·6	23·1
Port and inland water	32	21900	271000		
Other transport and communication	45	29400	51000	5·13	214·7
Distributive trades	30	4300	53000	1·1	20·1
Administrative, financial and professional services	48	14200	199000	0·77	32·2
Miscellaneous services	16	4200	27000	0·75	12·7
Total	1086	409000	2944000		

*Expressed as stoppages per 100,000 employees.
†Expressed as days lost per 1,000 employees.
Figures calculated on June 1975 employment figures (*Department of Employment Gazette*, January 1976, pp. 20–3) and 'Evidence of Industrial stoppages in the UK' (*Gazette*, February 1976, pp. 115–26).

The coverage of the car industry could be justified by the extent of the stoppages in the area as revealed in Table 5.8. On all indicators there is no doubt that, coupled with the importance of the industry, the period in question was by no means peaceful and television coverage therefore appropriate. However, we are here concerned not with what was present in the bulletins but what was absent. In terms of stoppages and days lost in relation to levels of employment in the industry, shipbuilding demonstrates a more serious position, not reflected in the bulletins. Also largely absent were three sectors where the frequency of stoppages per worker was very prominent – other sections of the vehicle industry not reported (re SIC 'all other vehicles'), engineering, and, above all, coal mining.

Thus, these aspects are not conveyed by the relative absence of strike reports from these important areas of industry. However appropriate the coverage of certain sectors might be, the lack of coverage of other sectors results in an overall skewing of the picture given of disputes in industry. The 'time defence' which justifies the selection process means that those sectors covered are thrown into sharp relief – less time more evenly distributed would better reflect the actual incidence of these events overall. Add to this the fact that reporting of strikes and disputes involving stoppages of work accounts for 39·1 per cent of all industrial coverage, but only 20 per cent of the total number of industrial stories. Table 5.9a shows the close similarity between the three news services in this respect.

Table 5.9a Percentage strike reports and stories of total industrial coverage

Channel	Bulletin reports %	Stories %
ITN	39·7	20·8
BBC1	39·7	23·9
BBC2	39·5	23·2
All channels	39·1	20·7

And there is a further point. In the nature of the case, the relative duration of disputes as against many other stories, means they are reported more often. This is indicated by the fact that they receive coverage in a greater than average number of bulletins than other kinds of industrial news story (Table 5.9b).

The skewed nature of reporting of strikes is correspondingly amplified by the repetition of reports that the nature of disputes demands.

Table 5.9b Average number of bulletin reports per industrial news story

Channel	Bulletin reports per story	
ITN	3·6	Industrial non-dispute stories
BBC1	3·6	
BBC2	2·4	
ITN	8·7	Dispute stories
BBC1	7·5	
BBC2	5·3	

The greater figure for strike coverage on ITN shown in Table 5.9b arises from the fact that ITN covered a smaller total number of strike stories, 36 compared with 45 different strike stories covered by BBC1.[12] At the same time, the total number of ITN bulletins that report these stories is relatively high and is comparable with the total for 45 strike stories on BBC1 (ITN total bulletin reports 313; BBC1 total 339: see Table 5.6).

Whilst BBC1 covered stoppages in 11 different sectors, ITN covered 8 sectors, none of which are not also covered by BBC1. BBC1 reported strikes exclusively in shipbuilding, textiles and distributive trades. In addition, ITN reported fewer strikes in 4 sectors – engineering, road transport, sea transport and air transport – but reported more disputes on the docks than did BBC1.

BBC2 bulletins, within the five-month period, covered strikes in 8 different sectors with similar absences to ITN in shipbuilding, and distributive trades. BBC2 did however report a strike in the textile industry but did not cover the dispute in the aerospace industry at the British Aircraft Corporation factory, Herne airport, which was reported by both ITN and BBC1.

It can be argued that television news gives much the same picture of the industrial scene, and indeed the world at large, with respect to the tendency to report the same aspects of a particular story from one news service to another and in choosing those stories from the same restricted areas of possible choice. The case studies demonstrate the close similarity of content and treatment found in the reportage of particular stories by the news bulletins on the three channels.[13] There is close agreement between ITN and the BBC as to which parts of the industrial world shall be allowed through the selection 'gate'. Even if different stories are chosen they tend to be drawn from the same sectors and tend to relate to the same sort of events. For example, both ITN and BBC1 cover 8 strikes each in the motor vehicle industry. ITN bulletins carry reports of three disputes not covered by BBC1, whilst BBC1 reports a further two strikes exclusively, and shares coverage of one other with BBC2.

The result of this is not significantly to alter the overall amount of car industry coverage on either channel. And anyway, such exclusives tend to be reported in just one or two bulletins.

The selection of news stories between channels is as follows: of a total of 36 strike stories chosen by ITN, 8 are exclusive to that news service. These are three strikes in the car industry, at Vauxhall Ellesmere Port (January), the BLMC Cowley plant (against short time) and Fords, Dagenham (April); two strikes on the docks, at Manchester (April) and in Hull (April); a one-day strike by teachers in Buckinghamshire (February); a strike involving Avon County Council workers (February), and the firemen's dispute in May. Only two of these stories appear in as many as three bulletins per channel. The remainder are reported in one bulletin each.

BBC2 has only one story which does not appear on the other channels, a dispute involving NGA members at the *Peterborough Standard* newspaper (*News Extra*, 29 May).

The BBC news bulletins cover some 17 strikes that ITN did not report. All of them appeared on BBC1, whilst seven of these were covered on BBC2 as well, indicating a significant level of co-operation and sharing of resources between the two news services. (In many cases the scripted studio report and the film reports used in BBC1 and BBC2 bulletins, covering the same story, are identical.)

The ten stories exclusive to BBC1 were strikes at Fords, Swansea (April) and at Fords, Halewood (February); two strikes by textile workers in Lancashire (5 March) and in Manchester (11 April); a strike by Co-operative Society workers (9 March); the ending of a two-year strike by teachers in Redcar; a dustmen's strike in Southwark; the strike action taken by NUPE members in the private beds dispute at Christie Hospital in Manchester; a strike at the Plessey telecommunications factory in Liverpool (25 February), and at Swan Hunters Shipyard (14 January).

The seven stories covered by both BBC1 and 2 (but not ITN) were a strike at Jaguar, Coventry (8 and 9 May); the strike and demonstration of Lancashire textile workers (25 March); a London Transport strike in protest against football hooliganism (May); a Southampton dockers' strike (May); the strike of seamen caused by the introduction of a new ferry (May); a strike by Manchester airport ground staff (February), and the strike by maintenance engineers at Heathrow airport in May. As with the ITN exclusives, these stories received only 'light' coverage. In only two cases out of the fifteen is there a story included in more than two or three bulletins. Exclusive stories tend to be treated 'lightly' and not run in many bulletins, and exclusive strike stories are no different from other industrial stories in this respect.

There are 28 news stories (of a total of 54) which are shared by both BBC1 and ITN. BBC2 reported 24 of these, omitting the strikes at BAC Herne airport (February); at Vauxhall, Ellesmere Port (4 April); the Scottish teachers' strike (January); and the unofficial strike of London prison officers (1 April).

The individual differences between news channels do however occur within an overall pattern of similarity and are minor variations on a common theme.

In ten cases ITN carries more reports on a story common to all channels and BBC1 has more than ITN on ten other shared stories. As would be expected, BBC2 has less reports per news story. Its stories are included in less bulletins than either BBC1 or ITN. The only exception to this occurred during the ACTT strike and lock-out during the last week of May, where BBC2 carried reports in twelve bulletins between 22 and 30 May, of the strike itself, as against fifteen on BBC1 and three on ITN.

The stories covered by all three news services are those, not surprisingly in the light of the other evidence on close similarity, which are deemed of greater significance. In only three cases for ITN and in one case for BBC1 is the story coverage run in less than three bulletins. In only one further case for ITN does the story receive less than two days' coverage. The number of reports per story amongst this group of shared stories, is above the average for strike stories (Table 5.9c).

Table 5.9c Extent of coverage, strike stories covered by all channels

Channel news service	Av. no. of bulletin reports	Av. days of story coverage
ITN	12·3	6·3
BBC1	11·9	6·2
BBC2	6·4	4·8

'Principal disputes'

A useful review of the performance of television news covering the 'big story' amongst industrial disputes is provided by the Department of Employment survey of 'Principal Disputes' in 1975. The Department of Employment analysis selects those disputes which resulted in the most serious stoppages of work, recording a high number of working days lost, causing noteworthy disruption of production.

We have listed all those principal disputes as published by the Department of Employment, which took place during our recording period, and examined the extent and nature of the coverage they received in the news bulletins. The details of this analysis are contained in Table 5.10 and prompt the following observations.

Of 20 strikes selected by the Department of Employment analysis as 'principal disputes', television news covered 11. Ten strikes in this group are covered by all three news services (these 10 being amongst the 24 stories common to ITN, BBC1 and BBC2 dealt with above). An additional story, the strike against the imposition of short-time working at the Cowley car plant, is the only exclusive and was reported by ITN.

There is a significant level of agreement between news services. However, there is an interesting tendency in reporting principal disputes for ITN to report over a longer period of time and in a greater number of bulletins. ITN's journalistic criteria for an important dispute story thus correspond more closely to the DE's notions of principal disputes than do the BBC's. Only in the case of the London Transport strike in January and the Birmingham toolmakers' strike in February, does BBC1 give comparatively more coverage in terms of number of bulletins and days run.

Table 5.11 compares the incidence of stoppages recorded by the industry sector with the strike coverage selected by the TV news. The television news bulletins did not cover the major strike in the shipbuilding industry or in the other manufacturing sectors. The fact that 5 of the 11 strikes selected for reporting are in the car industry, out of the 7 deemed important by the Department of Employment, again speaks for television's attraction to this area of the economy in 1975.

The importance of the engineering industry, as reflected in the incidence of major stoppages, is clear – 6 principal disputes in the first 5 months of 1975. Only one of these was reported on television. Two of the engineering disputes, which were not reported, occurred in Coventry where disputes in the car industry were routinely covered. The absence of coverage is not then a result of geographical distance or inaccessibility nor the special nature of the subject. In fact engineering is reported only with respect to its implications for the motor

Table 5.10 Principal disputes—extent and nature of coverage

Principal disputes*	Month	SIC order	BBC1 Days reported	BBC1 No. of reports	BBC2 Days reported	BBC2 No. of reports	ITN Days reported	ITN No. of reports	Total Days reported	Total No. of reports
Sanctions in support of a pay claim imposed on certain jobs by workers at a Coventry machine tool factory led to the suspension of an employee for refusing to carry out his normal duties. Some 1,600 production workers then withdrew their labour on 2 January in protest. Work was resumed from the nightshift of 15 January following acceptance of an improved pay offer made by the company.	Jan.	7	—	—	—	—	—	—	—	—
At an Oxford car assembly plant, 250 engine tuners involved in a grading dispute withdrew their labour, initially on 6 January, and continuously from 9 January, after unsuccessful interim talks. At present classed as production workers, the tuners were demanding skilled status which could benefit them in forthcoming pay negotiations in which higher differentials for skilled workers would be an issue. About 12,000 body-plant workers who were laid off on 6 January worked normally thereafter and were able to maintain production at a high level. The tuners returned to work on 4 February, on union direction pending the outcome of an investigation into the status question initiated by the Advisory Conciliation and Arbitration Service (ACAS).	Jan./Feb.	11	16	35	9	18	19	41	20	94
In support of a demand for better protective measures against assaults by members of the public, and in token of sympathy, on the day of the funeral of a	Jan.	22	4	7	5	5	3	4	6	16

conductor who died as a result of such an incident, London Transport and Home Counties bus crews, depot staff, and sections of London Underground Railway Staff, stopped work for 24 hours on 29 January. Nearly 18,000 people were involved.										
Objection to the grading within the pay structure of a new automatic plating process introduced by a Coventry telecommunications equipment company led to a stoppage by 27 platers on 10 January; as a result over 1,700 production workers were laid off progressively during January and February. Following a meeting chaired by ACAS, Midland Region, work was resumed on 5 March to allow further negotiations on the details of a proposed settlement.	Jan./Feb.	9	—	—	—	—	—	—	—	—
Electricians employed by local authorities in various areas of Scotland withdrew their labour in support of a claim for wage parity with contracting electricians in the private sector. The stoppage, which began on 10 January, was still in progress at the end of February, when nearly 900 workers were involved, principally in the Glasgow area.	Jan./Feb.	27	5	6	1	1	3	3	7	10
A six-week stoppage by 70 crane drivers at a Birkenhead shipbuilding yard, in support of a claim for wage parity with boiler-making trades caused the progressive layoff during this period of 1,200 other workers. The claim was not conceded, but acceptance of a lump sum advance payment in respect of oiling and greasing time was followed by a return to work on 24 February.	Jan./Feb.	10	—	—	—	—	—	—	—	—

Table 5.10 continued

Principal disputes*	Month	SIC order	BBC1 Days reported	BBC1 No. of reports	BBC2 Days reported	BBC2 No. of reports	ITN Days reported	ITN No. of reports	Total Days reported	Total No. of reports
At a Birmingham car plant about 600 toolmakers stopped work on 10 February after rejecting a pay offer by the company in annual negotiations. Their action in support of a demand for an increase which would restore their traditional pay differential with production workers, resulted in the progressive layoff of over 4,000 other workers at the same establishment. The dispute was unresolved at the end of the month. At the same company's Oxford plant more than 800 warehouse workers withdrew their labour on 18 February in protest against security measures which led to allegations that an employee was followed home by a works policeman. The stoppage ended on 27 February.	Feb./Mar.	11	4	6	3	3	2	2	7	11
In a dispute with the management of a Stafford firm of electrical engineers over pay differentials between testers and other skilled workers, an offer based on a pay and grading exercise was rejected and 118 testers stopped work on 26 February in support of an across-the-board increase. As a result 1,200 production workers were laid off from 8 March. A resumption of work took place on 26 March after the testers had accepted a marginal graded increase.	Mar.	9	—	—	—	—	—	—	—	—
At a Bathgate, Scotland, truck and tractor assembly plant, 96 electricians and pipefitters withdrew their labour from 6 March in support of their union's refusal to accept a new joint wage and conditions	Mar.	11	—	—	—	—	—	—	—	—

agreement operative from 3 January, recently accepted by unions representing hourly paid workers. The issue was related to pay differentials between skilled technical staff and semi-skilled operatives. Work was restarted on 24 March after minor modifications had been made to the agreement. The stoppage had meanwhile caused the progressive layoff of 3,800 other workers.										
A stoppage of work by 9,000 dockers in the Port of London which began on 27 February continued throughout March. The dockers' objective was to secure more of the container handling work at inland depots retained by road haulage contractors. Intensive picketing by dockers involved counter-action by road haulage workers. A panel of investigation was set up by ACAS. The dockers voted overwhelmingly for a return to work on 7 April.	Feb./April	22	13	27	9	10	17	35	23	7
Drivers employed by Glasgow Corporation cleansing department returned to work on 14 April after a thirteen-week stoppage during which troops were used to start clearing the tons of refuse building up throughout the city. The drivers, holders of HGV licences, claimed parity with rates paid by private hauliers, and their stoppage was supported, progressively by drivers from other corporation services.	Jan./April	27	19	36	13	17	21	42	27	94
A seven-week stoppage by 160 male employees at an East Kilbride telephone and cable factory caused 1,000 women production workers to be laid off. The men were in dispute over a pay offer which the women, who had been awarded an additional increase towards equal pay, had accepted. Following a meeting with ACAS, Scotland, agreement was reached under a new offer and work was resumed on 14 April.	Feb./April	9	—	—	—	—	—	—	—	—

Table 5.10 continued

Principal disputes*	Month	SIC order	BBC1 Days reported	No. of reports	BBC2 Days reported	No. of reports	ITN Days reported	No. of reports	Total Days reported	No. of reports
At a Coventry engineering firm which supplies the motor industry, 700 clerical workers withdrew their labour on 18 April in support of a claim for more pay to close part of the gap between their earnings and those of the manual workers. This led to the progressive layoff of about 2,000 production workers at the same plant and caused over 12,000 workers to be laid off in the motor industry elsewhere due to lack of essential components. The stoppage was still in progress at the end of the month.	April/May	7	7	16	8	9	10	16	11	41
In protest against 150 of their colleagues being put on short time, 2,000 indirect workers at a Ford car plant withdrew their labour on 18 and 21 April in the first of a planned series of absences to coincide with the pattern of short-time working. As a result 1,300 production workers who were not on short time were laid off. The dispute had not been solved at the end of the month and the stoppage was repeated on 2 and 5 May.	April/May	11	—	—	—	—	2	3	2	3
Six hundred maintenance engineers at a Wolverhampton tyre and rubber plant returned to work on 14 April after a four-week stoppage which caused 3,900 production workers to be laid off. Settlement of the dispute, which was over weekend working and the maintenance of pay rates, followed acceptance of the company's offer to increase the basic rate of pay.	Mar./April	19	—	—	—	—	—	—	—	—

Most independent television companies resumed broadcasting on 30 May after being off the air for seven days. A three-day withdrawal of labour by about 2,000 technical staff in support of a claim for pay allegedly lost during the period of wage restraint in 1973 was followed by a lockout. The claim remained unresolved at the end of the month.	May	26	7	15	8	12	2	3	9	30
A stoppage by 70 doorhangers and welders which began on 23 April caused about 5,000 workers to be laid off at a Dagenham car plant. The manning dispute, which arose over management's proposal to reduce the number of doorhangers on each shift in the body plant, was still in progress at the end of the month.	April/May	11	4	5	1	1	2	2	5	8
At a Coventry engine plant 4,000 production workers withdrew their labour on 9 May in support of a pay claim for an increase in basic rates of £15 a week. As a result of their action 3,700 workers at the Company's Ryton factory were laid off during the second week of the stoppage. Work was resumed on 5 June, following acceptance of an interim offer of £8 a week which shop stewards intimated they would seek to improve.	May/June	11	16	37	15	24	14	31	19	92
About 4,600 workers employed by a Coventry agricultural machinery manufacturer stopped work on 1 May having failed to reach agreement over annual pay negotiations. During the stoppage, picketing and occupation prevented administrative staff from entering the main factory. An improved pay offer led to the decision, by a narrow majority of workers, to end the stoppage. Work was resumed on 16 June.	May/June	7	—	—	—	—	—	—	—	—

* Source for this column: *Department of Employment Gazette,* February–June 1975

Table 5.11 Principal disputes compared to TV reports

Industry	No. of stoppages recorded by Dept of Employment	No. of strikes reported by TV bulletins
Engineering	6	1
Shipbuilding	1	—
Motor vehicles	7	5
Other manufacturing	1	—
Transport and communication	2	2
Miscellaneous	1	1
Public administration	2	2
Total	20	11

industry. The one dispute selected by television news was the strike by clerical workers in April at the Dunlop components factory in Coventry. In this case there was a clear link with the car industry since it led to progressive layoffs of workers at BLMC. This aspect of the strike was predominant in reporting and was the news angle that lifted the story out of the obscurity of the engineering industry onto the TV screen. The remainder of the television reporting shows a selection that mirrors the Department of Employment's selection of significant industrial disputes.

A two-month comparison

It is possible to mount a critique of television news journalism in terms which question the basic assumptions of all current journalistic practice, whether broadcast or print. Much of the above is on this level. The absences noted, the angles used and the tendencies exhibited, are not specific to television. They also apply to the press.

Yet it is possible that a further sense of the assumptions used by television newsmen could be gained by comparing their output with that of their print colleagues. A second level of critique specific to television, but leaving aside basic questions as to the nature of news, could thus be created.

To attempt this, we have taken a two-month period, January and February of 1975, and effected a comparison by counting the number of disputes as registered by the Department of Employment and the number of times they were reported in the media, broadcasting and press. The sample for this research was all the nationally networked television news bulletins on all three channels for this period, five national newspapers (the *Daily Mirror*, the *Daily Express*, the

Daily Telegraph, the *Guardian* and *The Times*), and the general wire service abstract of the Press Association.

The Press Association was founded by a group of provincial newspaper editors in February 1870 to break the monopoly of the private telegraph companies – an action which led directly to the GPO taking over the telegraphic service.[14]

Today, in line with its initial *raison d'être*, the Press Association services local newspapers and broadcasting organisations, but also provides a full news service to the national press. It was the abstract of this full Press Association service, described as the 'General Wire Service', that was used as the basis of comparison with the television bulletins.

The Press Association General Service is limited by the amount of teleprinting it can generate in any one day. It operates between the hours of seven and midnight and in addition to its reporting, it carries hourly news headline summaries. It runs four minutes behind the main service of the Press Association. There are no other constraints. That is to say, in print terms, it does not have to balance copy against space or, in broadcasting terms, copy against time. The considerable fluctuations from day to day in the number of Press Association stories, therefore, were a reflection of the 'gatekeeping' of their reporters and editorial staff which is, as it were, more purely limited by base notions of what constitutes news than any of the takers of the service can afford themselves.

The sample of national newspapers was drawn up to account for a large sector of the readership and spanned the range of readerships by social class.[15] However, in order to highlight any possible contrast between the range of press and television reportage, the sample was skewed towards the 'quality' dailies, as such papers aim for both range and depth in their reporting. In addition, as described in Chapter 4, there exists a similarity between the profile of the television bulletins and that of the quality dailies by story category. But despite this tendency for the television news to favour the 'quality end' of the spectrum, we wished to examine whether its news values in the industrial area were actually closer to those of the popular press. For this sample, a count was made of dispute reports receiving bold type headline or any mention in excess of ½ column inch. Hence, single sentence mentions of disputes 'packaged' with other items were not included in this newspaper count.

The PA tape, in contrast to the newspapers who normally print only one report per edition and the television news which presents one report per bulletin, appears continuously over some 17 hours, often carrying reports, reminders and updates of the same story several times in one day. Thus in the PA analysis, only the presence of reports, regardless of the repetitions, was recorded.[16]

Reports of 'threatened' strike action, that subsequently materialised, ongoing developments and settlements of disputes, were also counted. But reports of work-to-rule, overtime bans and work-ins, were not included in the count, as these are not recorded by the Department of Employment statistics as industrial disputes under the stoppages of work heading. The Department of Employment record of strike activity could then be 'read' against the media record.

During January and February of 1975, the PA reported disputes in 13 out of the 22 sectors in which disputes were recorded, covering 10 sectors in January, 7 in February (Table 5.12). The press sample showed a broader range of 16 sectors over two months, 13 in January and 12 in February (Tables 5.13 and 5.14). The television, however, shows a much reduced range, covering disputes in 8 different sectors overall, 6 in January and 6 in February (Table 5.15).

Table 5.12 Press Association dispute reporting, January and February 1975

SIC	January		February		Total	
	Disputes	Reports	Disputes	Reports	Disputes	Reports
Mining	—	—	1	1	1	1
Coal and Petroleum products	1	1	—	—	1	1
Metal manufacture	—	—	1	1	1	1
Engineering	1	4	1	1	2	5
Shipbuilding	2	2	—	—	2	2
Vehicles	5	27	6	13	10	40
Textiles	1	1	—	—	1	1
Paper, printing and publishing	2	19	—	—	2	19
Construction	1	1	—	—	1	1
Transport and communications	5	11	8	30	12	41
Distributive trades	—	—	1	1	1	1
Administrative, financial, business and professional services	4	29	7	28	11	57
Miscellaneous	2	5	—	—	2	5
Total	24	100	25	75	47	175

N.B. Several of the disputes are reported in both January and February, thus total disputes over the two-month period in some sectors are not simply additions of the monthly dispute columns.

Table 5.13 Press coverage of disputes, January

SIC Sector	*Daily Express* Disputes	Reports	*Daily Mirror* Disputes	Reports	*Daily Telegraph* Disputes	Reports	*Guardian* Disputes	Reports	*Times* Disputes	Reports
Food, drink and tobacco					1	1				
Chemicals							1	1	1	2
Engineering					1	1	2	2		
Shipbuilding					1	1			1	1
Vehicles	2	13	2	18	2	20	1	17	2	22
Textiles	1	1					1	1	1	1
Paper, printing and publishing	2	7	2	3	2	15	2	15	2	12
Transport	4	11	3	6	5	14	5	12	4	13
Distributive									1	1
Administrative, financial and professional	3	6	2	2	6	11	10	16	7	12
Miscellaneous	1	1	1	1	1	1	1	1	1	4
Total	13	39	10	30	19	64	23	65	20	68

Table 5.14 Press coverage of disputes, February

SIC sector	*Daily Express* Disputes	*Daily Express* Reports	*Daily Mirror* Disputes	*Daily Mirror* Reports	*Daily Telegraph* Disputes	*Daily Telegraph* Reports	*Guardian* Disputes	*Guardian* Reports	*Times* Disputes	*Times* Reports
Coal mining					1	1				
Metal manufacturing	1	2	1	1	1	2	1	3	1	3
Engineering									1	1
Shipbuilding					3	3	1	1		
Vehicles	3	7	2	4	2	7	3	10	4	14
Paper, printing and publishing	1	1			3	5	3	4	3	4
Construction					1	1	1	1		
Gas, Electricity and Water							1	1		
Transport	3	20	5	17	6	31	8	35	6	26
Distributive					1	1	2	2	1	1
Administrative, financial and professional	3	6	1	1	4	10	8	17	9	11
Total	11	36	9	23	22	61	28	74	25	60

Table 5.15 TV coverage of disputes, January and Febuary

SIC sector	Dispute story	January ITN	BBC1	BBC2	Total Jan.	February ITN	BBC1	BBC2	Total Feb.	Total Two months
Metal Manufacture	Ebbw Vale Steel Plant					3	2	1	6	6
Engineering	Plessey, Liverpool	—		—		—	2	—	2	2
Shipbuilding	Swan Hunter, Shipbuilders	—	1	—	1	—	—	—	—	1
Aerospace	British Aircraft Corporation			—		2	2	—	5	5
Motor vehicles	BMLC, Cowley	38	33	16	87	3	3	2	8	95
Vehicles Motor vehicles	Vauxhall, Ellesmere Port	1	—	—	1	—	—	—	—	1
Motor vehicles	Ford, Halewood	—	—	—	—	—	2	—	2	2
Motor vehicles	BLMC, Castle Bromwich	—	—	—	—	2	4	3	9	9
Paper, printing and publishing	National Graphical Ass./Newspaper Proprietors Ass.	16	9	9	34	—	—	—	—	34
	NATSOPA/*Mirror* Newspapers	9	8	5	22	—	—	—	—	22
	London Transport, Bus Crew	3	6	3	12	1	1	2	4	16
Other transport and	British Airways, 'Shuttle' service clerks	—	—	—	—	14	8	6	28	28
communication	British Rail signalmen	—	—	—	—	17	15	7	39	39
and Port and inland water	Docks (mainly London) containerisation dispute	—	—	—	—	1	1	—	2	2
	Manchester airport 'Service Air'	—	—	—	—	—	2	1	3	3
Education,	Teachers, Redcar	—	1	—	1	—	—	—	—	1
Administrative, financial and	Buckinghamshire teachers		—	—		1	—	—	1	1
professional services	Scottish teachers	1	3	—	4	—	—	—	—	4
Public Administration	Glasgow Corporation Dustcart drivers	6	6	3	15	3	—	—	3	18
financial and professional	Glasgow Corporation Maintenance Electricians	1	4	1	6	—	—	—	—	6
services	Scottish Ambulance Officers	9	11	5	25	—	—	—	—	25
	S. Wales Hospital staffs (mainly Swansea)	—	—	—	—	8	3	1	12	12
	Avon County Council workers	—	—	—	—	1	—	—	1	1
	Total	84	82	42	208	56	45	23	125	333

Within the press sample, there were marked variations between the coverage of individual newspapers. The *Daily Mirror* reported disputes in 6 sectors, the *Daily Express* in 8. In contrast, the *Guardian* reported over 15 sectors, the *Daily Telegraph* 13, and *The Times* 12. Although the television news shows a similar profile by story category with the quality press, the range of industrial dispute reporting is more akin to the popular dailies.

The quality papers report disputes over a wide range of the economy, covering many other areas of industry eschewed by the popular press. The television coverage shared the 6 sectors reported in the *Daily Mirror*, only the BBC differing its inclusion of one report on shipbuilding (BBC1 21.00, 14 January) and two reports on

Table 5.16 Number of stoppages, January and February 1975

SIC	January 1975			February 1975		
	No. of stoppages	Stoppages in progress		No. of stoppages	Stoppages in progress	
	beginning in period	Workers involved	Working days lost	beginning in period	Workers involved	Working days lost
Agriculture, forestry, fishing	—	—	—	—	—	—
Coal mining	18	3,100	4,000	16	2,600	5,000
All other mining and quarrying	—	—	—	—	—	—
Food, drink and tobacco	5	300	2,000	12	1,100	8,000
Coal and petroleum products	—	500	6,000	—	—	—
Chemicals and allied industries	6	1,200	4,000	10	4,800	13,000
Metal manufacture	1	400	3,000	12	8,100	14,000
Engineering	29	12,700	70,000	48	17,800	70,000
Shipbuilding and marine engineering	8	9,000	68,000	9	6,400	41,000
Motor vehicles	5	13,900	29,000	15	21,100	67,000
Aerospace equipment	2	300	1,000	5	500	4,000
All other vehicles	1	—	1,000	—	—	—
Metal goods not elsewhere specified	10	3,300	16,000	12	1,400	24,000
Textiles	4	2,300	8,000	7	2,500	5,000
Clothing and footwear	—	200	1,000	5	1,000	7,000
Bricks, pottery, glass, cement, etc.	3	200	—	4	900	—
Timber, furniture, etc.	1	200	—	4	900	—
Paper, printing and publishing	3	1,800	3,000	4	200	1,000
All other manufacturing industries	2	2,000	10,000	2	100	4,000
Construction	8	900	5,000	21	2,620	26,000
Gas, electricity and water	2	300	1,000	1	100	1,000
Port and inland water	2	700	3,000	7	10,400	17,000
Other transport and communication	4	18,000	21,000	12	4,400	10,000
Distributive trades	5	2,000	29,000	8	400	4,000
Administrative, financial and professional services	8	2,300	20,000	13	5,300	58,000
Miscellaneous services	—	—	—	6	1,300	5,000
Total	127	75,700	306,000	230	93,300	381,000

engineering (BBC1, 17.45 and 21.00, 25 February). In contrast, the *Guardian* covered 7 additional sectors and the *Daily Telegraph* 5 sectors, which received no coverage in the television bulletins.

The number of working days lost through disputes, although an indicator of strike severity, does not necessarily have any effect on 'newsworthiness' for television, as we have shown above. This seems to be equally true of the other media (Table 5.16). In the case of the construction industry, where the Department of Employment recorded 8 major stoppages involving a total of 900 workers and 5,000 days lost in January, there was no coverage in either the National Press or TV.[17]

In the distributive trades, 5 major disputes began in January involving 2,000 workers, resulting in 29,000 working days lost. Whilst *The Times* and *Guardian* covered only one such dispute on one day, neither the PA nor the television carried any reports. Disputes in this sector accounted for 9·5 per cent of the total working days lost in January.

There are other examples where only light coverage was given to many sectors where a large number of working days were lost through disputes. For example, in the engineering industry, 77 disputes involving 30,500 workers lost 140,000 days during January and February – the highest recorded figure of any sector in the two-month period. The PA reported 2 of these disputes on 5 days. In the press sample there were only 3 reports concerning 2 disputes (2 in the *Guardian*, 1 in the *Daily Telegraph*). On the television one dispute was covered in 2 bulletins.[18]

The shipbuilding industry is similarly neglected by the news media. Here the Department of Employment recorded 17 major disputes involving 15,400 workers, resulting in 109,000 working days lost in January and February. The Press Association covered 2 disputes on one day each. Of the press sample, only the 'qualities' included reports, 6 in all covering 5 different disputes. The television bulletins reported none of these disputes, but the BBC included a brief mention of the settlement of a previous strike in its *Nine O'Clock News*, on 14 January. Shipbuilding accounted for 15·7 per cent of working days lost over the first 2 months of the year and received 1 per cent of dispute reports in both press sample and PA.

A consideration of those disputes that featured extensively in news coverage again indicates the absence of any direct relationship between the number of working days lost and the amount of coverage that disputes in any area of the economy will receive. A high proportion of working days lost in any dispute or any industry, does not in itself appear to constitute news.

In paper, printing and publishing there was extensive coverage of disputes in all three media. In the Press Association tapes the sector ranked fourth highest in terms of dispute coverage (19 items covering 2 disputes), representing 10·8 per cent of the total dispute coverage over 2 months, and 19 per cent in January. In the press sample this sector ranked fourth (67 reports covering 4 disputes) with 12·8 per cent of total dispute coverage for 2 months, 20 per cent in January.[19] In terms of the Department of Employment record of working days lost, this sector in fact ranked eighteenth with 1 per cent of the total in January, and 0·6 per cent over the two-month period.

Despite the rather arbitrary relationship between strike severity and news coverage, it remains that the three sectors given most prominence in the media all had a significant dispute record. But a careful consideration of the overall pattern of coverage reveals its highly mediated character.

Table 5.17 Press and TV coverage of particular economic sectors

SECTOR	DE Working days lost Rank	%	DE Workers involved Rank	%	P.A. Dispute items Rank	%	Press Dispute items Rank	%	TV Dispute items Rank	%
JAN										
Vehicles	3	10·1	2	18·76	2	27	1	33·4	1	41·5
Transport	5	7·8	1	24·7	3	11	3	21·5	3	7·5
Public admin.	6	6·5	7	3·0	1	29	2	17	2	24·1
FEB										
Vehicles	1	18·4	1	23·1	3	17·3	3	16·4	2	19·3
Public admin.	3	15·1	7	5·7	2	37·3	2	17·6	3	13·4
Transport	5	7·0	3	11·5	1	4·0	1	51	1	60·5

It can be seen in Table 5.17 that the press and television concentrated heavily on one particular sector of the economy in each of the two months. In January 'vehicle construction' was the top dispute sector; in February, transport. As we have said, disputes in these sectors did rank high in terms of both working days lost and workers involved, but not so high as to justify, of itself, their dominant position in the coverage.

Having given a particular priority to disputes in a sector, this priority does not appear to rise or fall in a simple relation to the working days lost or the number of workers involved in disputes in these sectors.

As working days lost is obviously a retrospective indicator, journalists might more readily make judgments about newsworthiness in terms of the number of workers involved in a dispute. However, the amount of coverage does not appear to be affected by either factor. In the case of Transport, there was a small increase in working days lost and a fall in the number of workers involved in disputes, from January to February. Television coverage of disputes in this sector, however, increased from 7·5 per cent of all dispute coverage in January to 60·5 per cent in February.

Converse, television coverage of disputes in the vehicle construction industry, fell from 41·5 per cent of all dispute items in January to 19·3 per cent in February. During the period, working days lost through disputes increased from 10·1 per cent in January to 18·4 per cent in February, as did the number of workers involved, these increasing from 18·7 per cent to 23·1 per cent. A similar inverse relationship exists for strike activity recorded by the Department of Employment and coverage within the press and Press Association.

This erratic shift in emphasis is the result of the highly specific focus upon chosen disputes within some sectors.This practice routinely occurs across both press and television, but it is particularly evident in the popular dailies and the television news bulletins because of their narrower total range of stories. 'Heavy' coverage becomes proportionately 'heavier'. There is a tendency to overstate contextually the significance of disputes isolated in this way.

The high proportion of television coverage in the motor vehicle industry is accounted for by one dispute – that of the engine tuners at the BLMC Cowley plant. In fact, one other motor vehicle dispute was reported that month, a strike at the Vauxhall plant at Ellesmere Port. This appeared exclusively on ITN (17.50, 13 January 1975). Coverage of the Cowley dispute thus accounted for 96·6 per cent of the total dispute coverage of the motor vehicle industry in January and 41 per cent of the total disputes covered in all industries in that month. As can be seen in Table 5.18 the heavy coverage given to this story is evidenced equally by all three channels, all bulletins. Emphasis on the Cowley dispute is also evident in the press. Cowley coverage accounted for 94·5 per cent of all dispute reports of the motor vehicle industry in January, and 31·8 per cent of the total of all dispute coverage.

In the Press Association tapes, Cowley reports accounted for 77·7 per cent of the coverage of disputes in the motor vehicle industry in January and 27 per cent of the total dispute coverage. As was

Table 5.18 January—coverage of Cowley disputes

Medium	Total of disputes recorded No. of disputes	No. of reports	No. of Cowley reports	% of all dispute reports
PA	27	100	21	21
Press				
Daily Express	13	39	11	28·2
Daily Mirror	10	30	17	56·6
Daily Telegraph	19	64	19	29·6
Guardian	23	65	17	27·4
Times	20	68	21	30·8
TV Bulletins				
BBC1 Lunchtime	6	16	10	62·5
BBC1 Early Evening	9	30	10	33·3
Nine O'Clock News	9	36	12(1)*	33·3
Newsday	7	17	8	47
News Extra	6	19	7	36·8
News Review	6	6	1	16·6
First Report	7	22	9	40·9
ITN Early Evening	8	27	13	48·1
News at Ten	5	35	16	45·7

* Figure 1 in brackets indicates a further bulletin (not normal transmission time) on BBC1.

indicated above, the wider range of coverage over all news categories in the PA is evidenced within the industrial category reports. Thus Cowley accounts for a smaller percentage of all such items in January although it receives comparable emphasis in terms of total dispute reports.

The 'quality' press shows a similar range to that of the Press Association whereas the range of the popular dailies is much narrower. Thus Cowley reports represent a greater proportion of the dispute coverage in the popular press as other stories are not so much in evidence. The *Daily Mirror*, for example, gave 56·6 per cent of its dispute coverage to Cowley.

This heavy emphasis is further influenced by the tendency to run only one or two items on most of those disputes covered. The popular press rarely carries updates on disputes, most of which as stated, appear only once.

The range of television coverage was narrow. Eleven dispute stories appear in January, the overall result being again a high proportion of Cowley reports. In the remainder of its coverage, the television news, generally for all channels, displays the tendency noted in the popular dailies, running only 1 or 2 items per dispute story before dropping it for another.

The heavy coverage given to one dispute in the car industry must be seen partly as a consequence of a journalism which describes the world via particular cases which are effectively set before us as a general model. News is produced as a set of discrete stories whose taken for granted 'newsworthiness' is the criterion of their 'natural' selection. Selected news events are often presented as implying significance for whole areas of industry not otherwise reported, and are more likely to receive prominence as news if they can be characterised or defined in a given frame of reference.

The strike by engine tuners at the BLMC Cowley plant, in January 1975, was newsworthy because it was 'a manifestly avoidable stoppage of production'; because of the role of BLMC in the economy not least as an exporter; and because of the general 'crisis' in the car industry. Such recognisable news angles could account for the dispute's extensive coverage in January.

Given that all these factors remained of great importance throughout 1975, the dramatic change in the amount of coverage of disputes in the motor vehicle industry in February, points to a peculiar inconsistency in the application of news values. It is almost as if the newsmen themselves were becoming bored with their own coverage, assuming a similar boredom in their audience and readership and in consequence cutting back.

The viewer of the television news might well have believed that in February strikes in the car industry ceased to appear as a major problem. Disputes in the car industry simply did not figure prominently in February. The television news covered 3 disputes; a strike at the Ford body plant, Halewood (28 February 1975, BBC1 21.00 and BBC2 12.55), a strike at the BLMC Castle Bromwich plant (21, 23, 24, 26 and 27 February in 4 bulletins on BBC1; 3 on BBC2; and 2 ITN bulletins); and on 3 February some reports of the Cowley tuners' strike which began in January (in 3 BBC1 bulletins, 2 BBC2 and 3 ITN bulletins).

Of the three disputes reported in the television news, the Castle Bromwich strike featured most prominently in all the media sampled. 'The strike has made 14,300 idle at three plants. . . . Production of the Mini and Jaguar lines has been halted. . . . The strike has cost British Leyland nearly £9 million in lost production at the rate of 400 Jaguars a week and 2,800 minis' (*Financial Times*, 28 February 1975, p. 12). Six hundred toolmakers were on strike for over three weeks.

The coverage this dispute received in no way reflected its similarity to the Cowley strike of the previous month. The Press Association devoted 21 per cent of its dispute coverage in January to Cowley and 6·8 per cent to Castle Bromwich in February. The quality papers devoted 28·9 per cent to Cowley in January and 14·8 per cent to Castle Bromwich in February. The popular papers devoted 40·5 per cent to Cowley in January and 11·8 per cent to Castle Bromwich in February. The television bulletins devoted 41·8 per cent to Cowley in January and 7·3 per cent to Castle Bromwich in February (Table 5.19).

Table 5.19 Media coverage of Castle Bromwich strike

Medium	Dispute reports Number	% of all disputes
PA Teletape	5	6·8
Press sample		
Daily Express	4	13·3
Daily Mirror	3	12·5
Daily Telegraph	4	6·5
Guardian	7	9·8
Times	9	16·0
TV		
BBC1 Lunchtime	—	—
BBC1 Early Evening	2	12·5
Nine O'Clock News	2	9·5
Newsday	2	14·3
News Extra	—	—
News Review	1	20·0
First Report	—	—
ITN Lunchtime	—	—
ITN Early Evening	1	4·5
News at Ten	1	4·5

A comparison of the potential news angles reveals one significant difference – the Prime Minister's speech on strikes in the car industry by then nearly two months old. The 'crisis' in the car industry had not disappeared in February and British Leyland's significance for the economy was presumably unchanged. It would seem that the labelling of the Cowley tuners' dispute by the Prime Minister was crucial for that event to emerge so prominently as news. As the case of the study of the Cowley strike demonstrates,[20] the Prime Minister's statement provided the media with a dominant view on which to hang the tuners' story.

In February, the focus changed, as we have suggested, not simply because the pattern of events changed. The disputes that received the most extensive coverage that month were mainly in transport – railway signalmen, London bus drivers and the British Airway's 'shuttle' strike.

The coverage of industrial accidents on TV news

Although one cannot know when or where industrial accidents are going to occur, it is possible to establish a base line in which the incidence of accidents may be ascertained across a range of industries. Here we may have a good working idea of which industries are relatively high risk so far as serious and/or fatal accidents are concerned. In HM Chief Inspector of Factories Annual Report 1974,[21] Appendix 5 provides information reported to SIC order 3–20 (including minimum list headings) which covers manufacturing and construction. The information brought together there gives, for 1974, total and fatal reported accidents and the incidence rates of

Table 5.20 Reported accidents 1974 by industry (Standard Industrial Classification 1968), with: (1) annual all accidents incidence rates, 1974 (per 1,000 employees at risk); (2) averages of fatal and group 1 incidence rates 1971–4 (per 100,000 employees at risk)

SIC	Total reported accidents		Incidence rates of rep. accid. per 1,000	Av. of incidence rates per 100,000	
	Total	Fatal		Fatal	Group 1
3 Food, drink and tobacco	24992	24	39·0	3·3	650
4 Coal and petroleum products	1718	3	55·6	10·0	610
5 Chemical and allied industries	9825	46	34·6	9·0	590
6 Metal manufacture	26685	53	70·4	14·2	1050
7 Mechanical engineering	25772	20	36·4	3·3	650
8 Instrument engineering	1473	1	15·0	0·8*	250
9 Electrical engineering	12547	3	20·6	0·8	320
10 Shipbuilding and marine engineering	8939	15	61·5	13·0	840
11 Vehicles	17371	10	30·9	2·2	460
12 Metal goods not elsewhere spec.	16949	16	35·8	3·1	670
13 Textiles	12447	9	26·4	2·7	500
14 Leather, leather goods and fur	745	1	19·0	3·1	400
15 Clothing and footwear	2699	2	7·0	0·4	110
16 Bricks, pottery, glass, cement, etc.	11474	16	45·5	7·2	790
17 Timber, furniture, etc.	6946	14	30·1	4·2	780
18 Paper, printing and publishing	10064	8	24·2	2·1	500
19 Other manufacturing industries	8444	3	30·6	2·6	510
20 Construction	—	—	—	—	—

* This is a correction to the published figure following consultation with the relevant department of employment—statistician.

reported accidents per 1,000 for that year. In addition, for the period 1971–4, averages of fatal and group 1 (i.e. the most serious for the person) accidents, are given per 100,000 employees at risk. A summary version of this table is reproduced here at SIC industrial order level (Table 5.20). Since this is the report of the factory inspectorate some industrial areas such as sea transport are not covered. The television news ran some stories in these areas as noted above.

For 1974, the three industries having the most fatal accidents were construction (166), metal manufacture (53), and chemical and allied industries (46). The industries with the highest incidence rates of reported accidents for the year 1974 were metal manufacture 70·4, shipbuilding 61·5, coal and petroleum products 55·6. For the period 1971–4 the industries with the highest fatality rates were construction 18·6, metal manufacture 14·2, and shipbuilding 13·0. For the same period, the industries with the highest group 1 accident rates were metal manufacture 1050, shipbuilding and marine engineering 840, and bricks, pottery, glass, cement, etc., 790. On all criteria therefore, metal manufacture and shipbuilding are the most dangerous industries to work in.

Table 5.21 shows TV coverage of accidents as they relate to manufacturing and construction. In the manufacturing and construction section of industry, 8 accident stories were reported – 6 on BBC1, 6 on ITN, and 2 on BBC2. BBC1 and ITN reported industrial accidents on seven days. The difference in the number of mentions between BBC1 and ITN is accounted for by two more mentions of the Flixborough explosion inquiry by ITN and the reference to diving accidents in the oil industry which BBC did not cover.

The eight stories straddle three SIC categories and half of them are located in chemical and allied industries. As it happens, the four accident stories constitute almost the total profile of the chemical industry so far as TV coverage is concerned. One of the stories is a foreign one relating to a plant explosion in Antwerp when six people were killed. Another explosion occurred in a Kent chemical plant when one person was killed and a case of workers being poisoned in the Gosport Cyanamid factory was reported. The Flixborough explosion took place in 1974 and is reflected in Table 5.21. The reports here relate to the findings of the Court of Inquiry.

It is of course entirely appropriate that the Flixborough story, which was by any definition an industrial accident of great magnitude, should be updated to include the Court of Inquiry findings. The 1974 Annual Report records that 'a total of 28 people working on the

Table 5.21 Industrial accident coverage on TV, January–May 1975, for manufacturing and construction

SIC order	Event		Coverage BBC1	BBC2	ITN
Oil and Petroleum	Fire on oil rig.	No. of days	1		1
	Oil industry	No. of mentions	1		1
	accident.	No. of days			1
	Cousteau criticism	No. of mentions			1
	on diving deaths.				
Total stories	2	No. of days	1		2
		No. of mentions	1		2
Chemical and allied industries	Flixborough				
	Inquiry into	No. of days	2	1	2
	explosion.	No. of mentions	3	2	5
	Kent chemical				
	factory explosion.	No. of days			1
		No. of mentions			1
	Gosport Cyanamid factory				
	workers poisoned by	No. of days	1	1	1
	gas leakage.	No. of mentions	3	1	3
	Antwerp, Belgium,				
	plant explosion fatalities.	No. of days	1		1
		No. of mentions	2		3
Total stories	4	No. of days	4	2	5
		No. of mentions	8	3	12
Construction	Portsmouth				
	Motorway bridge	No. of days	1		
	1 killed.	No. of mentions	1		
	Liverpool site	No. of days	1		
	1 killed.	No. of mentions	1		
Total stories	2	No. of days	2		
		No. of mentions	2		
Grand total	8	No. of days	7	2	7
		No. of mentions	11	3	14

site at the time (i.e. 1 June 1974), were killed and 36 others were injured. Outside the works injuries and dangers were widespread but no one was killed. However, 53 people were recorded as casualties by the casualty bureau and hundreds more suffered relatively minor injuries. Some 1,821 houses and 167 shops in the surrounding area suffered damage' (p. 15). However, it is perhaps necessary to point out that the chemical industry is by no means the one with the highest incidence rates as can be seen from Table 5.20.

The construction industry and the coal and petroleum products grouping, are both examples of industries with high accident rates, especially the first as Table 5.20 makes plain. These two industries

Table 5.22 Principle wage settlements reported in 1975 (January–31 May 1975) and extent of TV coverage

Wage settlements (Source: *Department of Employment Gazette,* February–June 1975.)				Extent of coverage								
Date of agreement	Operative date	Industry or undertaking and district	Brief details of change	SIC order	BBC 1 Reports	BBC 1 Days	BBC2 Reports	BBC2 Days	ITN Reports	ITN Days	Total Reports	Total Days
8 January	12 January	Gas supply—GB	Increase of 21p an hour (inclusive of consolidation of threshold payments of 11p an hour) to all adult workers, with proportional amounts for young workers.	21	–	–	–	–	–	–	–	–
27 January	3 February	Building—GB	Increases (inclusive of consolidation of £4·10 a week threshold payments) in standard rates of £5 for craftsmen, of £4·40 for labourers, with proportional amounts for young workers together with increases of 40p or 20p a week in guaranteed bonus and the introduction of a Joint Board Supplement of £2·60 a week for craftsmen and £2·20 for labourers.									
	30 June	Building—GB	Weekly increases: standard rates: craft operatives, male £3, female £8 or £8·20. Labourers, male £2·40, female £3·40 to £6·80. Joint Board Supplement increased by £2·40 for craft operatives and £2 for labourers. Guaranteed minimum bonus increased as follows: craft operatives 60p; labourers 80p.	20	2	1	2	1	2	1	6	1
27 January	3 February	Civil engineering construction—GB	Increases (inclusive of consolidation of £4·40 a week threshold payments) in standard rates of 12·5p an hour for craftsmen, of 11p for general operatives, together with increases of 40p or 20p a week in guaranteed bonus and the introduction of a Joint Board Supplement of £2·60 or £2·20 a week respectively, with proportional amounts for trainees and young workers.									
	30 June	Civil engineering construction—GB	Increases for craftsmen of 7·5p an hour in standard rates, of £2·40 a week in Joint Board Supplement and £0·60 a week in guaranteed bonus. General operatives' increases —6p, £2 and 80p.									

30 January	20 February	Laundering (Wages Council)—GB	Increases (inclusive of threshold payments of 6p an hour) in general minimum time rates of 13·87p an hour for adult male and female workers 19 and over with varying amounts for young workers.	26	–	–	–	–	–	–	–	–
20 February	First full pay week commencing on or after 31 March	Cast stone and cast concrete products—England and Wales	Increase of 8·375p an hour in minimum basic rates.	16	–	–	–	–	–	–	–	–
24 February	24 February	Retail meat trade—England and Wales	Increases (inclusive of consolidation of £4·40 a week threshold payments) of varying amounts according to occupation and area for workers 21 and over with proportional amounts for young workers.	23	–	–	–	–	–	–	–	–
1 March	First pay day in March	Footwear manufacture—UK (except East Lancs and the Fylde Coast)	Increases in minimum day wage rates of £1·375 a week for men 19 and over, of £1·875 for women 19 and over with proportional amounts for young workers.	12	–	–	–	–	–	–	–	–
3 March	1 March	Coalmining—GB	National standard weekly rates increased (inclusive of consolidation of £4·40 threshold payments) by amounts ranging from £9 to £16 according to occupation for adult workers, with proportional amounts for young workers, together with the introduction of a national production bonus scheme.	2	32	15	19	13	27	12	78	17
	First pay week in July	Coalmining—GB	No national production bonus was payable during the third quarter of 1975: this involved reductions in minimum entitlement of £2·90 a week for adults and £1·95 for juveniles.									
5 March	26 March	Dressmaking and women's light clothing (Wages Council—England and Wales)	Increases in general minimum time rates of 9p an hour for men 21 and over, of 11·25p for women 18 and over with varying amounts for late entrants, learners and young workers.	15	–	–	–	–	–	–	–	–
	31 March		Increases in general minimum time rates of 3p an hour for all workers with proportional amounts for learners and young workers.									
12 March	10 March	Retail distribution . Co-op societies—GB	Increase of £2·50 a week for male and female workers 21 and over, with proportional amounts for young workers.	23	–	–	–	–	–	–	–	–
12 March	14 April	Seed crushing, compound and provender manufacture—UK	Increase in basic rates of £6·11 a week for men and £6·61 for women 18 and over inclusive of consolidation of £3 from existing payments.	3	–	–	–	–	–	–	–	–

Table 5.22 continued

Wage settlements (Source: *Department of Employment Gazette*, February–June 1975.)				Extent of coverage								
Date of agreement	Operative date	Industry or undertaking and district	Brief details of change	SIC order	BBC 1		BBC2		ITN		Total	
					Reports	Days	Reports	Days	Reports	Days	Reports	Days
19 March	14 April	Clothing manufacture—GB	Increases in general minimum time rates and yield levels of 6p an hour for all workers.	15	–	–	–	–	–	–	–	–
25 March	1 January	Post Office—UK	Increases of varying amounts following revision of pay scales for manipulative grades (postmen, postmen higher grade, telegraphists, telephonists and postal officers).	22	2	1	3	2	3	2	8	2
10 April	First full pay period commencing on or after 1 March	Rubber manufacture—GB	Increases in minimum earnings levels (inclusive of threshold payments of £4·40 a week) of £7·50 a week for men, of £8·25 for women, with proportional amounts for young workers.	19	–	–	–	–	–	–	–	–
5 May	1 April	Vehicle building—UK	Increase in minimum wage rates of 8·75p an hour for adult workers with proportional amounts for young workers.	11	–	–	–	–	–	–	–	–
8 May	5 May	Retail distribution: Co-op societies—GB	Increases (inclusive of consolidation of threshold payments of £4·40 a week) of varying amounts, according to occupation, for general distributive workers.	23	–	–	–	–	–	–	–	–
14 May	26 May	Engineering—UK	Increases in national minimum time rates of £4 a week for skilled men, of £3·20 for unskilled men and £3·25 for women, with proportional amounts for young workers.	71819	2	1	–	–	2	1	4	1
	24 November	Engineering—UK	Increases in national minimum rates of £4 a week for skilled men, of £3·25 for unskilled men and of £4·25 for women (thus giving parity with unskilled male rates).									
16 May	5 May	Cotton spinning and weaving—Lancashire, Cheshire, Yorkshire and Derbyshire	Increase in current wage rates of 12·5 per cent.	13	–	–	–	–	–	–	–	–

20 May	26 May	Shipbuilding and ship repairing—UK	Increases in national minimum time rates of £4 for skilled workers, £3·62 for semi-skilled workers and £3·25 for unskilled workers, with proportional amounts for young workers.	10	1	1	1	1	2	1	4	1
	Beginning of pay week containing 24 November	Shipbuilding and ship repairing—UK	Increases in national minimum time rates of £4 a week for adult skilled workers, of £3·63 for semi-skilled and £3·25 for unskilled workers, with proportional amounts for young workers.									
22 May	12 March	Electricity supply—GB	Increases in salaries ranging from £375·50 to £507·50 a year for adult workers, with proportional amounts for young workers.	21	16	9	11	9	9	6	36	13
23 May	Beginning of first full pay week in June	Motor vehicle retail and repair—UK	Increases in minimum rates ranging from 4·5p to 10·25p an hour according to occupation for adult workers, with proportional amounts for young workers and apprentices together with a restructuring of grades.	23	–	–	–	–	–	–	–	–
27 May	14 June	Post Office—UK	Cost of living supplement of 5 per cent on basic rates for manipulative grades.									
	18 July	Post Office—UK	Further non-enhanceable cost of living supplement of 2 per cent on basic rates for manipulative grades.									
	First full pay week following 15 August	Post Office—UK	Further non-enhanceable cost of living supplement of 1 per cent of national basic rates for manipulative grades.	22	See ref. above 25 March		See ref. above 25 March		See ref. above 25 March		See ref. above 25 March	
	First full pay week following 12 September	Post Office—UK	Further non-enhanceable cost of living supplement of 1 per cent of national basic rates for manipulative grades.									
	First full pay week following 17 October	Post Office—UK	Further non-enhanceable cost of living supplement of 1 per cent of national basic rates for manipulative grades.									
	First full pay week following 14 November	Post Office—UK	Further non-enhanceable supplement of 1 per cent of national basic rates for manipulative grades.									
27 May	First full pay week following 12 December	Post Office—UK	Further non-enhanceable cost of living supplement of 2 per cent of national basic rates for manipulative grades.	3	–	–	–	–	–	–	–	–
30 May	2 June	Cocoa, chocolate and sugar confectionery manufacture—GB	Increases of £4·35 a week for full-time men and women.									

account for the remaining four stories, two a piece. At the same time it can be seen that shipbuilding and marine engineering, metal manufacture and bricks, pottery, glass and cement, which have high accident rates on several of the criteria noted, are not reported at all. Given that the area of industrial accidents is a significant aspect of industrial life and work experience, the TV coverage is somewhat limited. With the particular exception of Flixborough, there is no clear rationale to why some events are covered and not others. It appears, as it were, to be a hit and miss affair with no clear evidence of systematic coverage in this area. In so far as any rationale does appear to operate, it is linked with the coverage of disaster stories rather than with any routine concern with accidents in industry as a general social issue.

The general case for a more pronounced concern with accidents at work is detailed in *Safety and Health at Work* – the report of the Robens Committee 1970–2. (Cmnd 5034, HMSO, 1972).

And a further point can be made. General industrial health is particularly significant in the light of the high health hazards run in certain industries such as mining. Needless to say, the more general issue of industrial health receives little or no coverage on the television news.

Pay claims and wage settlements

Table 5.22 lists the principal wage settlements concluded between 1 January and 31 May 1975 and records the extent to which these settlements (and in some cases the negotiations preceding them) were covered on the three television channels. The six industrial sectors reported were mining, engineering, shipbuilding, the construction industry, electricity supply and the postal services. Reports were carried by all news services in each case, with the one exception that BBC2 did not mention the engineering pay deal. Although there were some twelve settlements in Table 5.22 that were not reported at all, it can be seen that these included a number of groups that, as far as the economy on the whole is concerned, are somewhat peripheral. It would have been surprising to have had a report on the seed crushing, compound and provender manufacture pay settlement for instance. But in enough of the major industry wide agreements are reported to make this an area of routine news coverage.

In four of the cases, the focus was on the settlement itself, referred to on one or at the most two days. These are clearly quite different in

Table 5.23 Additional coverage of wage claims and settlements (January–May 1975)

SIC	BBC1		BBC2		ITN		Total	
	Rep.	Days	Rep.	Days	Rep.	Days	Rep.	Days
01 NUAAW claim	—	—	—	—	1	1	1	1
06 Steelworkers' claim	1	1	4	3	2	1	7	3
22 Seamen's claim	5	3	2	2	3	2	10	4
25 Teachers' EIS claim and settlement Scotland	4	3	—	—	2	2	6	3
25 University teachers' claim	3	2	2	1	3	1	8	2
25 Teachers' claim	1	1	1	1	—	—	2	1
25 NHS doctors' pay rise	2	1	1	1	2	2	5	2
25 Junior doctors' and GPs' claim	4	3	4	3	6	5	14	5
25 NHS Radiographers' pay deal	2	1	1	1	1	1	4	1
25 Dentists' pay deal	—	—	—	—	1	1	1	1
27 Armed forces' pay rise	3	2	3	2	2	1	8	3
27 Police pay claim	3	2	—	—	—	—	3	2
27 Civil Service pay	1	1	—	—	1	1	2	1
27 MPs' pay rise	—	—	—	—	1	1	1	1

emphasis to the mining industry and electricity supply industry claims which were followed over an extended period. The recent history of the mining industry in connection with its pay claim and strike that brought down the last Conservative government, is well known. The 1975 pay claim was defined as critical in relation to the future of the Social Contract and the eventual settlement provided the occasion for inquests and speculation on the likely efficacy of government pay policy. The power workers' claim was also seen, by the television correspondents, as a potential challenge to the government's policy, particularly because of the union's concern with the erosion of differentials brought about by flat rate wage increases. Both mining and electricity supply are examples of industries which can have a direct effect on the rest of the productive system in the event of industrial action and we believe the closer monitoring of these industries reflected that awareness and gave something of a dramatic character to reporting on the subject.

Our argument that pay claims and settlements are routinely reported is strengthened by further evidence of two kinds. First, a number of pay settlements not listed by the Department of Employment's analysis (see Table 5.23) were also reported. These related to teachers, radiographers, dentists, the armed forces and members of Parliament. Second, a number of claims which were actually settled after the period of our study were reported (see

Table 5.23). These concerned agricultural workers, steel workers, railwaymen, university teachers, doctors, policemen, seamen, and civil servants. In the case of the railwaymen, the coverage was extensive and included the signalmen's strike. Apart from national wage settlements, with which we have dealt here, there were other more localised claims (related to particular firms or geographical localities) that were reported as in the case of London docks and the Chrysler plant in Coventry.

All of this leads to the conclusion that pay deals and claims were indeed subject to regular scrutiny. The relationship of pay to the government's management of the economy and the health of their Social Contract may be seen as an important organising device for handling industrial and economic news; in fact for the majority of these stories the dominant view was often simply whether they were settlements within the Social Contract. The rival industrial correspondents did not always agree on this, as our case study in Volume 2 will show.

Journalistic criteria

The question may be put: if there is no clear relationship between indicators of real activity and industrial news coverage, what light does this throw on news values? At once the caveat should be entered that in the case of wage claims and settlements the Social Contract was used as the basis of reasonably systematic coverage. In relation to the other indicators, however – employment magnitudes, strike statistics and accident rates – there is no consistent matching between them and the events reported on television industrial news. The information given by television on these events is, in a statistical sense at least, unrepresentative. This itself is an important conclusion about the skewed nature of industrial reporting in relation to specified criteria.

The response may be made: news is about the extraordinary not the ordinary, about the 'significant' events not the mundane ones. This response has been partially considered in our discussion of principal disputes. To this we would add a further comment. It is not difficult for this response to operate as a circular argument. The significant is what is covered. The insignificant is what is not covered. It is quite a good protective device from a debating point of view since the concept of significant absences is ruled out by definition. It is also possible for this response to serve as an underpinning for a view of journalistic criteria which treats them as an arcane mystery known only to the practitioners of the craft. At the intuitive level it is all about having a 'nose' for a good story.

As an information service, there is something to be said for covering routine matters systematically and that this itself could constitute a journalistic criterion if policy governing news was so disposed. Moreover, as a matter of empirical analysis, the appeal to the value of significance, does not result in a random collection of stories, the anarchic product, as it were, of competing journalists and news services. What emerges is a pattern of coverage which in terms of the industries, the stories covered and the number of reports, reveals great similarities between BBC and ITN. It is like looking at the synoptic gospels, and facing the almost irresistible question: what is their common source?

Are there then principles of selection which may be detected and at least by implication, suggest something about prevailing journalistic practices? Recall the dominant coverage allocated to the car industry, transport and communications and public administration. The first of these is the pre-eminent example of mass-production industry. It may be used to summarise what are held to be the problems of production in an advanced industrial society: strike-prone workers who, despite high wages, are not content; the cycle of prosperity and depression always more dramatic with a consumer product in a mass market; the competitiveness of the international market and the relevance of this for the balance of payments; the relation of government to industry as it pertains to financial aid, control structure and the promotion of industrial efficiency. In a phrase, the car industry may be said to embody a concern with the principle of industrial (even social) survival in a society exposed to the stiff winds of competition and inflation.

The emphasis on transport and communications, and public administration, reveals, we suggest, another criterion. It is a concern for the inconvenienced consumer of goods and services. A strike that grounds aircraft is highly inconvenient to the holidaymakers and businessmen, a railway strike is very troublesome to the commuter, a doctors' work-to-rule or hospital workers' boycotting of private patients is distressing to the consumer of health services and a strike of dustcart drivers is a growing difficulty for the consumer wishing to dispose of his unconsumed leftovers.

These are two of the essential journalistic criteria in the industrial area, which are embedded into, and structure, the news on television.

They have to do with unscheduled interruptions to production processes and consumption patterns. Given this emphasis it is difficult to structure news in a way that does not implicitly, at least, blame those groups or individuals who precipitate action that, in one way or another, is defined as 'disruptive'. This structuring often demands a search for the 'disruptive' element, which is exacerbated by the lack of historical perspective – an element of news presentation that often results in a somewhat arbitrary allocation of blame for the disruption. The other side of the coin, the concern with a particular kind of social order, is revealed in the pre-occupation with the 'social contract' in its many ramifications. Thus we would deny that the constraints of bulletin duration, technical limitations, manpower, programme budgets, geographical and other access difficulties, result in a haphazard picture of industrial life. The journalistic criteria outlined above and other elements not mentioned, result in a coherent frame for the reporting of industry. The contours of coverage never deviate from this frame.

Given these working assumptions about industrial news values, we would be hard put to demonstrate any 'bias' at all.

6 TRADES UNIONS AND THE MEDIA

We have already argued that the news coverage which we have examined stands in a highly mediated relationship with the events which are the substance of its content. We wanted to better understand how the newsrooms were fed this content in much the same way as we felt the need to sensitise ourselves to the processes of news production. The literature allows two views of the producers' function. In one model the production process is seen as being haphazard with a fair degree of producer autonomy, and in another as the result of a series of binding professional, economic, technical and political constraints.[1] But either of these views of producers ignores the attitudes and strategies of those they are dealing with. This is of particular importance in the newsrooms. Since this need to qualify the research tradition was not fully conceived of at the design stage but emerged late in the first phase of the project, it was not then possible to establish contact with the total range of all parties used by and using the medium in connection with industrial life. Faced with a choice, but knowing that within broadcasting the hostility of trade unions was known to be regarded as a real constraint on coverage, we decided to concentrate on the unions' formal relationship with the medium. In particular, we wanted to assess how the structure and development of individual unions may affect this relationship. There may exist a general difference in the attitudes and approach of supposedly middle-class and working-class unions to the media in general. This involves the familiar distinction between white- and blue-collar workers.

In fact, this distinction makes less and less sense in the post-war period. As a change-around in the occupational structure of the British economy occurred, more and more trade unionists were to be found in so called white-collar unions. At the present time a minority of the labour force are involved in manufacturing, mining and extraction, the traditional blue-collar occupation. The changing division of labour has meant that from the 1960s most workers are actually in the supposed white-collar unions. Thus, the distinction is now relatively meaningless – in terms of criteria such as income, status and life chances; yet it remains of use for analytic and historical reasons, and major differences in union character along this axis in the 1970s can be expected.

Perhaps a more important distinction would be between élite (vertical) and non-élite (horizontal) unions. Some unions, e.g. NALGO, have top management and low-paid workers in the same union. These vertical unions often started as organisations of professionals but have now unionised vast numbers of low-paid workers. Indeed in a period of pay restraint and effective cuts, the majority of workers can be expected to be involved in tactics such as strikes which are more traditionally associated with manual or blue-collar workers. This heuristic distinction between white- and blue-collar workers is used here merely to examine whether differences towards media fall along this historical and cultural axis or along other criteria.

We sought to account for any differences by examining the inputs of trade unions in relation to the coverage which they received. Such inputs could be assessed by comparisons of the number of press conferences given, direct contacts with newsdesks, and the deployment of union resources on public relations, press and research back-up. We sought to gather such information as a background to our own observations and analysis of the 22 weeks of coverage which we had recorded, and to collect impressions on the general attitudes of trade unions to the media and in particular, the trade unionists' contribution to the debate on the future development of the media.

The range of trade unions covered broad sectors of the economy with variations in trade union composition and membership. Social research has been carried out in this area, notably by Lockwood in his study *The Black-coated Worker.*[2] This initiated debate around the possibility of a convergence between working-class and middle-class attitudes and lifestyles. In the light of this debate interviews were sought with union officials of both broad categories of unions.

The 'blue-collar' group includes interviews with the AUEW (Amalgamated Union of Engineering Workers), the GMWU (General and Municipal Workers Union), the TGWU (Transport and General Workers Union). Any differences within this the 'blue-collar' group relating to their histories and to their position as organisers in either the public or private sector of industry were noted. To accommodate this dimension the group was expanded to include the NUR (National Union of Railwaymen), ASLEF (Association of Steam Locomotive Engineers and Firemen), and NATSOPA (National Society of Operative Printers, Graphical and Media Personnel).

The corresponding group for white-collar unions included BALPA (British Airline Pilots Association), ASTMS (Association of Scientific, Technical and Managerial Staffs), recruiting chiefly salaried staffs including technicians, foremen and administrative personnel, the POEU (Post Office Engineering Union) and NALGO (National and Local Government Officers Association).

With this group, apart from the basic questions we were raising, the concern was to establish whether the very rapid growth of white-collar unionism throughout the 1950s and 1960s had altered their relationship with the media. The peculiar development of unions such as NALGO, from being essentially associations of professional staff to becoming fully fledged trade unions, was of interest here.

A further possible course for investigation was that certain of the white-collar/public sector unions might have a particular relationship with the media in as much as their members were defined, at least historically, as fulfilling a 'valuable' public service. To accommodate this aspect the range of white-collar/public sector unions was expanded to include the NUT (National Union of Teachers), NUPE (National Union of Public Employees, who recruit mainly hospital ancillary staff), and COHSE (Confederation of Health Service Employees, who recruit mainly nurses).

It should be said that the analytic categories of white-collar/blue-collar and of public sector/private sector, around which the sample of trade unions was organised, are not precise. This is so because the unions themselves overlap in their recruitment. ASTMS, for example, overlaps in some areas with NALGO, as at present in the campaign to organise university staffs. ASTMS, is also now attempting to recruit hospital staff and doctors. The Transport and General Workers Union, which has popular connotations as perhaps a classic blue-collar union, has in reality a large clerical and administrative section. It is actually in some competition with unions such as ASTMS for the recruitment of technicians and in 1969 changed the title of its clerical and administrative sections to the Association of Clerical, Technical and Supervisory Staffs.[3]

The vertical/horizontal distinction poses additional problems for the simple white-collar/blue-collar division. However, the main purpose of this research was not to draw precise distinctions between the composition of different trade unions. Rather it was essential that the main areas of the economy were covered and that a broad overview was given of those unions which received the most sustained coverage on the news in the period of our research.

There were obvious limits to our sample of trade unionists, not least that we were able to interview only one or two people from organisations which comprised many thousands of members. The interviews were designed to be open-ended to encompass the widest possible range of responses. However, the first part of the interview was structured to include the following three areas.

First, the extent to which the particular trade union organises its coverage to fit the requirements of the media to achieve maximum effect; second, how the development of the particular union and any changing organisational or membership patterns might affect its input to the media. This is a sensitive area since it may impinge upon conflicts and tensions which currently exist within the union. The questions were formulated with this in mind and the hope that the union representatives might feel confident enough to point to any effects on their coverage which may result from such tensions.

The third area was concerned with the general attitudes to the media existing in the trade unions. Here the interviewer would ask the interviewee if he could recall past coverage of other unions and how it compared with that of his own. These questions were intended to initiate a discussion on the general relationship between the media and the trade unions. Such discussions would involve any criticisms of the media and would attempt also to ascertain whether the officials believed any pattern to exist in the coverage of industrial affairs.

The social and economic context

The context for our research was necessarily set by the immediate history of the whole union movement especially as the inception of economic crisis from the late 1960s onwards and the growth of heavy and sustained unemployment has challenged the thesis that white-collar workers necessarily enjoy a superior position in the labour market. The progressive erosion of real wages for both white- and blue-collar workers, and the cuts in public spending, are of importance. Although wage increases are commonly attributed as the main cause of inflation, it is the case that, in the last ten years, average 'real' earnings have increased at well below the rate of inflation and since 1972 they have actually decreased as the rate of inflation continued to rise. A study of Inland Revenue annual reports and reporting of employment figures by R. Bacon and W. Eltis concluded that: 'the average living standard in these terms has risen only 1·3 per cent a year and since 1972, it has actually fallen.'[4]

At the same time unemployment in the 1970s has remained, except for the brief period at the end of 1973, higher than even the peak of unemployment reached in the 1960s.

In addition, the atmosphere of industrial life was changed in the period of Conservative government between 1970–4 by a series of political and economic confrontations. These included a series of strikes against government policies of wage restraint, the most notable of these being that of the coal miners in 1973–4 which led to the Conservative government calling an election which they lost. In addition, throughout the period from the late 1960s to the fall of the Conservatives in 1974, the trade union movement resisted attempts by both Labour and Conservative governments to limit their power by legislative means. In particular, the Conservative government's Industrial Relations Act did much to inject an atmosphere of political confrontation into industrial life.[5]

Commenting on these developments, J. Westergaard and H. Resler, in *Class in a Capitalist Society* note that:

> Demands for higher wages may be taken to the point where they challenge the continued viability of private capital. That is exactly what business today fears: and many recent wage claims have been of that order. They acquire political significance, moreover, precisely because business relies on the state to defeat them. That in itself is liable to turn 'economism' into something more; at least where there is a stubborn, though patchy, foundation of quasi-socialist counter-ideology within the labour movement to effect the translation. There were hints to point that way, when, for example, miners voted (against an insecure Labour government) to withhold support from any public policy of restraint on wages, 'so long as the capitalist private profit-making character of British society remains unaltered'; or when nurses in the course of a dispute about pay, withdrew their labour from private wards in public hospitals in the hope of ending private fee-paid treatment and

the associated privileges of consultants within a health service supposedly dedicated to equality of service for all. These were only straws in the wind. But there were others, in the late 1960s and early 1970s which suggested that demands arising in the first instance out of issues of employment, pay and prices could take a twist to question the principles of societal dominance by property and market.[6]

These changes may in part be measured by the increase in the number of disputes recorded per year. Since the period 1952–9, when the incidence of disputes remained fairly constant, there has been an increase in both the number of disputes and their duration, this latter an important indicator of the intensity of conflict. Between 1968 and 1973, strikes lengthened to an average of nearly seven days compared with only two and a half days in the earlier 1960s.[7] During this period trade unions increased their membership by over 1½ million between 1968 and 1974.[8] At the end of this period the total stood at 11,755,000.

The most profound differences distinguished between the two types of unions lie in the varied social compositions of their membership and in the periods in which they have undergone their most rapid growth.[9] This could be expected to be reflected in their attitudes to and relationships with, the media.

The blue-collar unions have a history of long and bitter conflict with the established press. Such a conflict emerged in the earliest, formative years of the British Labour movement. From the early nineteenth century the Labour movement engaged in attempts to set up its own press against an established press which was seen as openly hostile to its own interests. One such paper, *The Charter*, commented in 1839 that:

> The newspaper press, daily and weekly, is the property of capitalists who have embarked upon purely commercial principles, and with the purpose of making it contributory to their own personal and pecuniary interests. It is the course that is profitable, therefore, and not the course that is just that necessarily secures their preference.[10]

The history of the antagonism of the established press towards trade unionism throughout most of the nineteenth century is well documented.[11] The most notable response to this ongoing hostility was the establishment of the *Daily Herald* in 1911.

The hostility of the press during the First World War was reserved for those workers who went on strike, particularly if the strike was seen as damaging to the war effort.[12] The inter-war years saw little change in this mutual distrust. It is no accident that amongst the events surrounding the General Strike in 1926 one was the NATSOPA chapter of the *Daily Mail* refusing to print an anti-strike editorial. The Second World War saw the accelerated culmination of a period of rapid growth for blue-collar unions which began in 1918. During this period the TGWU rose from less than 700,000 to over a million members, and the Amalgamated Engineering Union more than doubled its size from 334,000 in 1939 to 825,000 in 1943. Overall membership increased by two-thirds.[13] Pelling argues that the unions gained not only in membership but also in prestige. But did this in any way affect the bitter press legacy? And did the introduction of television and its emergence as the front-runner national medium of mass communication, affect trade union hostility towards the press?

Basically it seemed that the hostility still existed. For example, Lord Cooper of the MGWU, during the strike at Pilkingtons in 1970, said, 'I am sick and tired of Press and TV reports based on information which is wrong and false. . . . The organs of publicity do not contribute to better relations in industry . . . damaged the name of the union and made things worse for us' (*Financial Times*, 18 May 1970).[14] But have attitudes changed since?

The MGWU is the third largest union in Britain with a membership of, at present, around 880,000, covering ten major industries including glass, chemicals and local authorities.

The MGWU had two full-time officers concerned with the press, broadcasting and public relations activities. This did not seem a very high proportion of a full-time staff of 550. There was a definite tendency throughout the whole interview the National Industrial Officer gave us for the interviewee to regard the media as meaning primarily the press. While the MGWU subscribed to a news agency service which provides them with press cuttings, there is no regular or sustained monitoring of television or radio output. Neither of the two press officers had apparently any direct or regular contact with the television newsdesks. When asked specifically about television coverage the interviewee tended to discount it. He argued that there was unlikely to be any in-depth coverage on the television because of the technical problems of giving such coverage and the shortage of time available. Second, he pointed out that he believed television coverage to be London-orientated and that most of the members in his union were in the north.

This attitude was later confirmed in a discussion on the effects of media coverage when the interviewee commented that he felt local newspapers to be the crucial area in which the union should concentrate its efforts. Also, in reply to the question 'How would you publicise a key change in policy?' the interviewee replied that he would call a press conference and that it would be timed at 11.30 a.m. 'in order to catch the evening editions of the papers'. Such timing might mean that they would miss any possible inclusion in the lunchtime news bulletins but, it emerged later, that this was not something that was normally taken into account.[15] The union did not take external professional advice on the planning of coverage or on personal presentation and interview strategies. There was no equivalent here of the public relations and publicity operations mounted by large corporations or, for example, the current campaign being mounted by the Confederation of British Industry.[16]

Rather, a key element in the attitude of the MGWU, later found to characterise a number of the blue-collar unions, was that they expected to be approached by the media rather than to initiate contact themselves. The MGWU interviewee gave a number of examples where second-hand and garbled versions of what the union had done and what its representatives had said had appeared in the press. The comment which accompanied these examples was that 'no one came and asked us what was really happening'.

A fairly low level of formal input to the media existed. There was a tendency to concentrate on the press rather than on television, and in the normal course of events the MGWU expected to be approached rather than to initiate any substantial contact with the media.

We found some of these features typified the approach of other blue-collar unions, notably the Amalgamated Union of Engineering Workers, the Transport and General Workers Union and the Association of Steam Locomotive Engineers and Firemen. The AUEW is the second largest trade union in the country with 1,400,000 members concentrated in light engineering, machine tools, vehicle production and shipbuilding. The TGWU is the largest union with around 1,800,000 members covering a broad range of industry including construction, public transport and vehicle building. ASLEF, the union of locomotive drivers, is a smaller union with a membership of 30,000.

In AUEW a National Executive Officer was interviewed who informed us that the central union organisation employed one public relations officer, who, in his words, 'issues occasional press statements'. In addition, a regular broadsheet on union policy, written by the General Secretary, is circulated to the media. This interviewee was also mainly concerned with the press, but it should be said that this particular officer had recently been the subject of a personalised press campaign which was critical of him.

In discussion the interviewee drew attention to the difference between one of the white-collar unions and his own. He mentioned Clive Jenkins and ASTMS as being particularly developed in the area of public relations and added his belief that 'most unions are not like this, they simply go along on their own way and wait for the media to come to them'.

The interview with the Head Research Officer of the TGWU confirmed the general pattern of low input so far established. This union has no publicity or public relations department, there is no single official who deals exclusively with either. They too tended to wait for the media to approach them and when this occurred, they directed them to those responsible for the relevant area. The Research Officer acted as a back-stop for any inquiries or approaches which fell into the miscellaneous category.

When issues of importance arise in the TGWU the General Secretary calls a press conference and releases are made through the Universal News Service (UNS) and the Press Association news services, and separate releases are made to the television and individual newspaper correspondents. However, this is an ad hoc procedure and no formal structure exists for a sustained organisation of coverage. In contrast to the MGWU the TGWU discuss broadcasting and the press as being of equal importance.

With the Association of Steam Locomotive Engineers and Firemen, all publicity and public relations goes through the General Secretary and his assistant who were interviewed. It is perhaps understandable that there should be no separate publicity department, since the union is fairly small and recruits within a limited range in a single industry. Notwithstanding this, the attitudes to media input were similar to those already outlined. There was a definite tendency to wait for approaches from the media and very few press conferences were called. There was a concentration on the press as the most important medium. On particular occasions, the union would contact the PA but no use was made of UNS. The significance of this latter news service is that it enables the organisations that subscribe to it to put a full and unedited statement through the wire service which it operates. Its use is one indicator of the union's publicity awareness.[17]

The general pattern, which has so far been established, did not hold for the National Union of Railwaymen. This union recruits most of the railway workers and has around 200,000 members. It has a full-time publicity department of five people plus research assistants and secretaries. The General Secretary, in an interview, stated that this department had been developed in a conscious attempt to 'exploit and use the media'. In addition, he commented that much of this new interventionist approach to both press and television had emerged since his appointment and he made clear his belief that the union should 'bend over backwards to accommodate the media'.

In addition to an extensive use of the normal channels of access, the NUR initiated direct contacts with those sections of the media which they sought to actively change. For example, the General Secretary referred to a campaign in which they had persistently hectored the editor of the *Daily Telegraph* to achieve what they saw as a more balanced account of a recent NUR pay claim. This kind of interventionist approach was found to be much more generally characteristic of the white-collar unions.

The enormous growth of white-collar unions since the Second World War, and the characteristics of white-collar workers, have been the subject of considerable social research. One area which has received attention is how the attitudes and perceptions of such workers affect their motivation to join trade unions. For example, in 1954 Strauss argued that: 'White collar workers join unions, not because they reject their middle class aspirations, but because they see unionism as a better way of obtaining them.'[18]

Strauss argued that if white-collar unions were to be successful in recruiting members, then they would have to adopt middle-class characteristics. This involved distinguishing themselves sharply from working-class unions, both in their general appearance as 'professional associations' and in the relationship which they have with their employers. However, as Blackburn argues in his later study *Union Character and Social Class*,[19] in practice there may exist some conflict between this aim of achieving a respectable and middle-class appearance and the potential effectiveness of the organisation as a trade union. This is more so if the achievement of the respectable image involves the rejection of the strike weapon or other forms of militancy in negotiations with employers. White-collar unions may thus exist in a precarious balance between respectability and effectiveness, and Blackburn argues the key to this balance is in the attitude of the employers. It has been subsequently argued that the potential for such a balance to be sustained and for a degree of equanimity to exist between employers and white-collar workers, turns upon the more general developments of the social and economic system. J. A. Banks, in his recent work on trade unionism, draws attention to this crucial interaction between unions and the system of social and economic relations within which they exist.

> The emphasis of trade unionism as a social movement, that is to say, draws attention to the collective endeavour on the part of trade unionists, within their separate unions and in co-operation between unions, to bring about some significant change in their situation and even more broadly, so to alter the economic and social systems in which they find themselves that they take another form, regarded as more desirable. . . . Thus trade unionism cannot be understood merely in terms of trade union participation in an industrial relations system, because the nature of that system itself experiences modification to the degree that trade unions are successful in their efforts to modify it.[20]

Indeed, it can be argued that this historical difference between blue- and white-collar workers, whereby it is assumed that the white-collar unions are uninterested in traditional union bargaining strategies, is now undermined by the numerical preponderance of white-collar workers in the economy. This means, despite their predilections for status rather than income, that their very size makes them inevitable targets of any income policy. The fact that a large proportion of these are now state employees contributes to this tendency, especially as in a period of economic difficulty there is always a campaign for cutbacks in public expenditure.

The development of the economic system in Britain through the period of relative affluence in the 1950s to the economic crisis of the 1970s is important since it was within this period that the white-collar unions experienced their most rapid growth.

Some researchers accounted for the earlier stages of the white-collar union growth as simply a phenomenon of the expansion of the public sector and the increase in white-collar employment in general. As late as 1966, Bain argued that:

> Since 1948 the absolute amount of white collar unionism has increased greatly. This has prompted many people to speak of a boom in white collar unionism. Such people are suffering from a growth illusion, which results from considering changes in union membership in isolation from changes in the labour force. In real terms this membership boom is non-existent. In spite of the phenomenal growth of some white collar unions, white collar unionism in general had done little more than keep abreast of the increasing white collar labour force, and the density of white collar unionism has not increased significantly in the post-war period.[21]

Bain's thesis, while tenable for the 1950s, does not hold thereafter. For example, in 1961 NALGO, which is the largest white-collar union recruiting in the public sector, had 273,600 members. By 1973, it had 498,000 members which represents an increase of over 90 per cent, while local authority employment in this period had risen by only 53 per cent.[22] In fact, the situation now is that the growth of the public sector is being curtailed, while union membership continues to rise. Other white-collar unions report a similarly rapid growth. The National Union of Public Employees has doubled its membership in the last nine years and stands in 1975 at 507,000 members. The Association of Scientific, Technical and Managerial staffs has had perhaps the most spectacular growth of all, since the founder union had a membership of only 18,000 in 1961, while ASTMS in 1975 had a membership of 350,000 and was growing at the rate of 1,000 per week. It is clear then that other explanations are needed for this growth than the relative increase in the white-collar working population.

Roberts, Loveridge and Gennard, in their study *The Reluctant Militants*, outline the structural changes which have undermined the position of some groups of white-collar workers. Commenting on technicians in industry, they note that:

> The position of the technician has been affected by other factors apart from a decline in relative pay. The trend towards larger and larger business units under the force of competition and the search for economic stability has led to increasing

> bureaucracy and remoteness. Although technicians have been aware of this trend, the personal relationship with management has prevented vigorous protest. Management, on the other hand, has increasingly come to see technicians as part of the new mass labour force, as a homogeneous group, little different from the manual workers on the pay-roll. Like manual workers, technicians feel they are becoming mere numbers on the pay-roll computer tapes.[23]

The decline in relative pay which is mentioned here has been of particular significance for employees in the public sector. The press officer of NALGO commented, when interviewed, that public employees had been affected severely by the pay policies of the government throughout the 1960s, where wage increases were kept at one stage to 3 per cent a year. How has this affected how these unions seek to handle the media?

The NUR's interventionist approach to the media characterised all of the white-collar unions spoken to, but with a view to assisting recruitment as much as for public image making or public communication with the membership. COHSE, for instance, is a relatively small union with a membership of 100,000 yet it maintains a full-time press officer and fairly high and regular input into the media. We interviewed the press officer and the General Secretary. In the 1974 dispute over nurses' pay the union set up a duty room with special responsibilities for dealing with the media. In this period daily press conferences were held and daily releases were made to the press and television, at both national and regional level, and through the PA news service. At times when they are not in dispute they issue releases through the PA service about twice a week, but they do not use the Universal News Service since they regard it as being too expensive. Since the last dispute this union has gone to some lengths to develop personal contacts with industrial correspondents and news reporters. But this effort to establish contacts must be seen in relation to its recruitment policy.

The General Secretary of COHSE described the potential members of his union as 'an oilfield to be tapped'. The account which COHSE published of the 1974 nurses' pay campaign begins with the record of how many members this campaign drew to the union:

> This is the record of the nurses' pay campaign during the summer of 1974, a time when morale and confidence had sunk

> so low that for the first time ever, many thousands of nurses reluctantly resorted to a campaign of industrial action to underline the necessity of an urgent, comprehensive remedy to their appallingly low pay. The campaign of the 'reluctant militants' was undertaken and led by C.O.H.S.E. and supported by many other nurses. That support was dramatically illustrated by a 26% increase in C.O.H.S.E.'s nurse membership to more than 95,000 in less than two months.[24]

Very similar attitudes and levels of input to the media are to be found in other public sector unions, notably in the National and Local Government Officers Association. NALGO is a much larger union than COHSE, with a current membership of around 625,000. It has a publicity department of 17, two of whom are engaged full time on press/broadcasting work. NALGO subscribes to the Universal News Service and in addition makes a frequent use of the Press Association service, sending out two to three press statements per week. This union has also initiated a series of public criticisms of the media and they take every opportunity to make these known; for example, in their submission to the Royal Commission on the Press and the Annan Committee on the Future of Broadcasting.

The importance of NALGO, at the outset essentially a staff association for local government officers, is that many of the most radical changes which have occurred in white-collar unionism are highlighted by its history. This development is also reflected by the changes which have occurred in the internal structure of the union's publicity department. Initially this department was organised very largely around the aims and interests of the local authorities who employed NALGO members. The publicity department was called the public relations department and was essentially engaged in public relations work for local authorities and, to a lesser extent, in presenting NALGO as a respectable middle-class association.[25]

Towards the middle of the 1960s, however, there was a definite change towards using this department more exclusively for the purposes of NALGO's own publicity. Its title was changed from the public relations department to the publicity department. As in the case of COHSE, this move was primarily the result of the union's growth, recruitment policy and the use of the media as a means of communication with members and potential members. At the same time NALGO's definition of itself was changing from a close identification with the employers, essentially as a staff association, to being much more like a trade union. For example, it had joined the Trade Union Congress in 1964 and had its first official strike in 1970. Thus, although NALGO has had from its very early stages a relatively high level of contact with the media, there has been in recent years a thorough transformation both in the quality of this input and the self-image projected by the organisation.

The high level of organised publicity was found to exist in other public sector unions, notably the National Union of Public Employees, which is a fairly large union of 507,000 members. In interview the Assistant General Secretary commented that his union was experiencing a change in its media image. It had been involved in recent years in a series of protracted disputes most notably over the 'pay beds' issue, where it had come out strongly against private practice in the Health Service. In the course of these disputes the Union had six people working full time at publicity and public relations. When not in dispute, fewer officials work in the area, but it would still normally send out two to three press statements per week and uses both the PA and UNS news services. In addition, it organises press conferences on specific issues and frequently complains about the coverage it receives. The Assistant General Secretary commented to us that he had recently been on the phone every day to newsdesks and copy-takers, for both the television and the press.

ASTMS, which has had perhaps the most rapid and spectacular growth of all the white-collar unions, developed very early the use of publicity for recruitment purposes. A series of mergers of major companies in the mid-1960s had resulted in a large number of technical staff being made redundant. ASTMS organised a recruitment campaign on this issue and in 1967 carried a full-page advertisement in *The Times*, headed 'Management has Decided it Does Not Like the Colour of Your Eyes'.[26]

This theme has recently been taken up by NALGO, who are organising a campaign against the proposed local government cuts, and are using slogans such as 'The Sack Race' and 'Solo Players Can Get Dealt a Lousy Hand'. The interview with the General Secretary of ASTMS illustrated the extremely high level of resources which this union devotes to publicity and public relations work. There are six full-time workers and press conferences are held every ten days. ASTMS does not use the Press Association news service as the General Secretary believes it to be too slow. Instead, the union has direct telex links with newsrooms and news correspondents. In addition the General Secretary placed great emphasis on the role of television – his press conferences are held in the morning and are aimed at *First Report*. The publicity operation of ASTMS is deliberately tailored to what they see as media's requirements and tasks. The General Secretary gave the example that if they were involved in a dispute and found out that the wife of someone on strike had had twins, then the union would send a twin pram along with the strike pay!

The General Secretary saw the key to the very rapid growth of his union first in publicity and second in the political education of existing members. The publicity is used to present the idea of what he called the 'new unionism'. By this is meant a unionism which caters for the new middle classes, technicians and managerial staff.

The National Union of Teachers, which has 250,000 members, takes a close interest in the wider issues of educational development. Therefore it is less concerned about media input as a recruitment strategy and, in part, has developed its approach to publicity to increase its effectiveness in public debate about education. The General Secretary, in interview, outlined the extensive publicity and public relations operation which they mount. The NUT employs nine people full-time in this area and the department in which they work is distinct from the research department and has been operating, in some form or another, for over ten years. The general pattern in seeking coverage is that statements and decisions on NUT policy are formulated by committees within the union and are then handed to the publicity department for distribution through the Press Association news service. On average, such statements are issued between two and three times each week. Exceptional decisions merit press conferences, which are held early in the morning to encompass the needs of both television and the press. On occasions when they are seeking the right of reply or are disturbed by the nature of particular coverage, they will initiate direct contact with the newsdesks involved. The General Secretary's comment, which summarises their attitude, is that they seek 'as much coverage where they can, when they can'.

The Post Office Engineering Union does not typically receive very much publicity, and recruits from one skilled section within a single industry. For this reason the POEU does not engage in the same level of publicity operations as the other unions in the white-collar group, but in some respects its attitudes and general approach to publicity is similar. The union has 127,000 members, and contacts with television and press are organised by the Assistant General Secretary (whom we interviewed) and one other person. There is no separate publicity or public relations department, but, in the words of the interviewee, they send out a steady stream of press statements and documents. The Assistant General Secretary had worked at one time on *Reynolds News* and his particular interest was with the press, but the union did take a considerable interest in television and had sent their ex-General Secretary on a 'How to Present Yourself on TV' course and had recently formulated a specific complaint to the BBC about coverage of them in the *Money Programme*.

In understanding the relationship of the POEU with the media, it should be remembered that they are, in part, a broadcasting union and that they organise a section of the labour force that is very rarely in dispute. The only significant attention that they had ever received from the media, was because of a one-day stoppage in 1969 when they had blacked out some sections of the commercial television network. This had been their only strike for approximately twenty years. Although they were very sensitive to the general practices of the media and their potential for communication, they reasoned that since they did not personally receive very much coverage, there was no point in spending money on publicity and PR work except when their particular interests were involved. Thus most of their efforts were directed towards issues such as government spending on the Post Office and campaigning against the existence of cartels amongst the producers of communications equipment, as well as taking part in the debate about broadcasting. They too presented evidence to the Annan Committee.

So far it has emerged that there are apparently considerable differences between unions in terms of the relative effort which they put into contacting and using the media, and in particular, between the inputs of white- and blue-collar unions. There are, however, some qualifications which should be made to this general picture. First, it might be the case that individuals who are employed in publicity and public relations work have a professional interest in arguing for the scale and value of their own operations. Second, it must be said that there is no formal measure, whether it be the enumeration of press conferences, the use of news services or the allocation of personnel, that can adequately encompass a union's approach to the media. For example, the allocation of personnel in some unions tends to be fairly fluid, and who is working on what is subject to some variation. We have already commented that attitudes as to the value of press services, such as the Press Association and the

Universal News Service, vary widely amongst trade unionists. Thus a simple count of the use or non-use of these services by a union is not an adequate guide to the acuteness of its interest in media input.

Notwithstanding this, there did emerge a general pattern of difference in the approach to the media between white- and blue-collar unions with the single blue-collar exception of the National Union of Railwaymen.

Union structure and media presentation

Do any severe divisions inside unions as to how publicity operations might be used, or over what kind of self-image the unions sought to present, exist? Tensions and conflicts over such issues might be expected to most affect those unions which were experiencing a rapid growth and consequent changes in their internal organisation. This is a politically sensitive area within which to inquire. The inadequacies of a brief interview with individuals from the top of the union hierarchy became, in this particular area of inquiry, immediately apparent. The effect of questions which related to tensions within the union, was to oblige some union officials, perhaps not unnaturally, to put on political hats. For example, questions as to the restrictions which might exist from the union's end were often met with general statements about the need to encompass the opinions of everyone in the union – that this was the job of the General Secretary, etc. Many union officials thus felt obliged to play down the idea of any tensions, or conflicts and instead to stress the consensual unity of their organisation.

We established closer research contact with one of our sample, the National and Local Government Officers Association, when we were invited to give a series of papers to union weekend schools. This enabled us to obtain some insight into the complex interaction which exists between the internal processes of the union and the definitions and descriptions which are offered by the television news producers. We offer one example as a qualification to the tendency in media research to see the processing of news into tight formulae, packages and definitions as being the result only of the routine practices of the producers of news, irrespective of questions as to how autonomous they are.

In the evidence given by NALGO to the Royal Commission on the Press, the union focused attention upon what it saw as the media's fascination with the effects of disputes and the 'disruption' caused by industrial action. It argued that the heavy emphasis on these aspects of disputes led to a serious neglect of the underlying causes of the conflict and the motives and reasons of those on strike, which might make their actions appear rational.

The evidence is essentially a case study of a dispute in which NALGO members were involved and was widely quoted to us by other trade unionists as an example of press bias:

> A brief analysis of newspaper coverage of a simple trade union dispute illustrates quite starkly the effects of a simplified news process. The dispute concerned up to 200 N.A.L.G.O. members employed by the British Waterways Board who took limited official strike action in October 1974 to back a pay claim following an unsatisfactory final offer by the employers. The background to the claim and action was fairly complex, involving an erosion of real incomes, the effects of reorganization, relativities with other groups, traditions of non-militancy, inadequate working conditions. Newspaper reports collected at N.A.L.G.O. Headquarters, formed a fully representative sample of the total coverage. The cuttings amounted to 800 column inches. Analysis showed the following:

	Column Inches
1. Descriptions of the effects of the strike, progress of strike, progress of the negotiations	715
2. Reference of the claim to arbitration	45
3. Settlement of the claim	25
4. Discussion of the background to the claim, conditions in the service affecting the workers, etc. (i.e. why action was necessary in the first place)[27]	15

In the light of this extensive and critical approach which NALGO had developed, it was surprising to find cases in which NALGO officials not only accepted the interpretive framework of the television news interviewers but actually also accepted the dominant view of the story. In one interview a NALGO district official appeared in the context of a threatened refusal to work by NALGO members on the EEC Referendum day. In the introduction to the interview the newscaster set the agenda in a fairly typical way, referring to the 'effects' of the dispute. This is exactly the approach which had been so bitterly criticised by NALGO the previous year.

The full text is as follows:

Holmes: 'Well all sorts of people oppose this referendum but not the trade unions, yet it may be trade unionists who prevent voting taking place on Thursday in parts of South Wales where local government workers have been asked by their union, NALGO, not to do any extra work that day. The local branch has already said that its members won't help to man polling stations or with the count; but in South Glamorgan, NALGO members say they will co-operate. Four other counties have yet to make up their minds. Patrick Hannan asked one of the NALGO officials if they weren't worried about the effect their action might have upon the referendum.'

Interviewee: (Name plus caption – NALGO District Official) 'Of course we are concerned but let's face it, many trade unions take action which can be of very . . . major concern to many people in this country and we are not doing anything which we can regard as being of any less concern than what some other, many other, trade unions have also done in their particular negotiations.'

P. Hannan: 'What do you anticipate will be the effect of the action?'

Interviewee: 'Well the effect will be that if there is sufficient support from NALGO members, backed by other trade unions in South Wales, then the first effect will be that if there is insufficient manning arrangements for the polling stations on Thursday, then it could well be that many polling stations will simply not open in South Wales with the result that the whole Referendum will have to be done all over again.' (BBC2, 23.35, 2 June 1975).

Typically, what is completely absent from the interview is any mention of what the dispute is about. It is normal practice for media interviewers to set the agenda of interviews around stereotypical interpretive frameworks, union activities included. If we compare the above interview with one taken from our analysis of news coverage of the car industry, a number of differences emerge in the response of the interviewee:

Harris (BBC): 'So in the context of Mr Wilson's speech, was this an avoidable strike? The Convenor at Cowley, Mr Doug Hobbs. . . .'

D. Hobbs: 'Well I believe that the management could have settled this claim almost immediately. At the final stage in procedure, the negotiators for the organisations said to the company to agree in principle to this claim and we would sort out any anomalies afterwards.'

Harris: 'How are the men at Cowley reacting to Mr Wilson's speech?'

D. Hobbs: 'Well personally I believe it was untimely. It was quite an empty contribution as far as I'm concerned. Since April of last year, we have worked consistently, all of us, to try and avoid any disputes whatsoever. In fact most of the production that has been lost, has been lost through either breakdowns or shortage of materials and we do recognise that British Leyland is, has got a problem, a cash-flow problem and we have worked very, very hard, both union and members, to try and eradicate this position.'

Harris: 'How does the prospect of no government cash for British Leyland strike you if the strike record doesn't improve?'

D. Hobbs: 'Well I believe the strike record has improved. Because we've got this situation today is unfortunate, but we must have a disputes procedure and at the end of the road take any necessary action, then quite frankly, I don't know where we will go, or we could alternatively go back to the days of the wild-cat stoppages, which I certainly don't want and I'm sure our members don't want.' (BBC1, 6 January 1975)

The district officer in the NALGO interview is made to accede to the newscaster's definition and then at the end of the interview actually extends it by adding a comment on the highest conceivable amount of disruption which the dispute could produce. Even allowing that the interview may have been cut, the extent to which the interviewee is prepared to discuss the effects of the dispute is still remarkable.

In the Cowley case, the media stereotype is similar in some ways to this in that it underlines strikes and disruptions as the major cause of the car industry's problems. However, in this particular instance, the dominant view within the frame is established by the systematic and frequent reference to a speech made by the Prime Minister on the car industry which included the phrase 'manifestly avoidable stoppages of production'.[28] This speech, as we argue below (Chapter 7), overwhelmed the coverage of the dispute, but the essential difference between Doug Hobbs on the one hand and the NALGO officer on the other can still be seen. By comparison with the NALGO interview, the Cowley shop steward rejected the interviewer's interpretive framework about strikes in general and the dominant view of Leyland's troubles as 'manifestly avoidable stoppages' in particular. A simple view would suggest that this refusal to accept the interviewer's point of view must always be in the trade unionist's interests, if the interviewer is questioning in a hostile way. But the NALGO example reveals that such a simple view is inadequate.

It became apparent, when we attended a conference of NALGO publicity officers and showed the tape of the NALGO interview to the delegates, that there existed a deep division within NALGO over the nature of the industrial and political policies which the organisation should pursue. In part, this represents a division between 'right' and 'left' wing politics, but it also relates to a wide range of issues arising from the organisation and structure of NALGO. It is perhaps natural that at the present time of economic crisis there might develop some conflicts over policy issues, particularly when there are fears that cuts in government expenditure might involve enforced redundancies in local government, which is NALGO's main area of recruitment. However, such conflicts are likely to be exacerbated in the case of NALGO, since it recruits vertically. This means that the possibility of redundancies does not affect all the membership with the same force.[29]

In addition, vertical recruitment, and the consequent high differentials in salaries (in the ratio of 1:12) between NALGO members, raise severe problems for the negotiation of wage settlements.[30]

A further problem is that traditionally a post in the union's hierarchy is, in many cases, concomitant with a high post in the employment hierarchy. This is frequently commented upon within NALGO.[31]

The conflicts which we have so far described within NALGO frequently focus around the self-image of the organisation and the descriptions and definitions offered of it by its members. One district official mentioned that it was possible to detect which 'side' in the internal debate a person was on, depending upon how they described the organisation – if they referred to it as the 'association', the traditional term, or if they called it simply 'the union'. Underlying these different terms is a range of beliefs and attitudes about how NALGO should act, how it should relate to other organisations, what tactics it should adopt and what public profile it should have.[32]

It seems likely to us that the comments made by the officer in the interview quoted above may be best understood in the context of the ongoing debate which we have outlined. In particular we would note the references to NALGO as a trade union and the explicit linking of the actions of NALGO with those of other unions. The final comments of the interviewee on tactics and the effect of the dispute begin also to make more sense. Rather than simply accepting the TV interviewer's framework, the officer was using this framework to stress NALGO's new trade union character, irrespective of his own political stance.

Similar comments were echoed within NALGO when a number of members advanced the view that more aggressive tactics would have to be employed if NALGO was to seriously oppose local government cuts and the possibility of enforced redundancies. In a sense then, the comments of the district officer in the interview, may, in part, be understood as one way of seeing what kind of organisation NALGO should be; how it should act and what kind of tactics it should adopt, against the background of the internal divisions within the organisation. Again it is probably of some significance that, as the BBC commentator noted, some branches had yet to make up their minds about how they should act in the dispute. The implications of this in terms of the internal political situation are further emphasised by the fact that the interviewee's comments were actually made against NALGO's official policy.

All of which is a caveat against a simple view of trade union preferences in coverage. Extensive coverage is not necessarily in the interests of the union as we shall show, nor is the acceptance of stereotyping necessarily perceived as being against its interests.

The complexity of the internal affairs of a union can contribute to the input although such complexities are unlikely to become the substance of the coverage.

Attitudes and orientations to the media

There is a remarkable similarity of attitudes on the part of all the trade unionists in our sample on the general nature of media coverage. There still existed a considerable level of hostility to the media.

The almost universal complaint from the trade unionists interviewed was that both television and press coverage focused upon unions exclusively as 'dispute' organisations. They argued that this obsession with strikes and 'disruption' on the part of the media, led to an almost total neglect of the wider role which trade unions play in the general organisation and administration of the economy. In addition, it was often commented that an advanced industrial economy could simply not survive without this constructive co-operation on the part of trade unions.[33]

A second complaint, frequently made, was that not only did media coverage focus mainly upon strikes but when it did so it concentrated upon the most trivial and sensational aspects of them. There were two qualifications to these criticisms. First, the unionists tended to favour the 'quality' press as against the 'popular' press in terms of giving the fairest coverage. Second, a distinction was often drawn between correspondents as 'individuals' and the organisations for which they worked. Our interviewees very rarely attributed malicious motives to industrial reporters, but tended rather to speak of the pressures upon journalists to produce within particular moulds.

These criticisms are highlighted in a formal complaint about an edition of the BBC's *Money Programme*, which was made recently by the Post Office Engineering Union to Broadcasting House.

> The general view expressed by most members was that it treated the telecommunications side of the Post Office in a trivial and superficial way and that as a result it would have been better if a separate programme had been produced dealing specifically with the complex issues of telecommunications. You will recall that most of the programme was devoted to the postal side of the business and whilst we all recognise there are many serious problems still to be resolved, the fleeting impression gained was that the programme was more interested in personalities than the financial, human and technical issues of maintaining this service. Whilst the telecommunication aspects of the programme were essentially correct, their presentation created a false impression particularly to those in the industry and failed to emphasise the contribution this union has made towards improving efficiency and financial stability. Can I mention a few issues which have a considerable financial bearing on telecommunications and which, one would have

thought, would have been included in a *Money Programme* series.

1. The union's campaign over a decade to break the 'ring' in the purchase of exchange equipment.
2. The union's campaign, again over many years, to persuade successive Governments to increase capital expenditure on telecommunications.
3. The union's successful campaign to substantially restore the capital expenditure cuts imposed by the former Chancellor of Exchequer, Tony Barber.
4. The union's contribution towards increasing productivity in the Post Office (probably one of the best records throughout industry).
5. The role of the union in the decision to purchase literally millions of pounds of exchange equipment over the next decade.
6. The union's campaign to still further improve Post Office efficiency by securing powers for the Post Office to enter the field of manufacturing.

These six points clearly indicate how trivial the production was for telecommunications and how totally inadequate the programme was in presenting an accurate picture of what is inevitably one of the largest (in financial terms) industries in the country.[34]

Beneath the generally heard criticisms of superficiality and triviality, some revealed more intensive critiques of the normal practices of media production. For example, it was pointed out to us that the media do not trivialise or sensationalise everything. Rather, coverage is organised into highly selective patterns which imply a definite way of seeing and understanding industrial life.

We refer to these patterns as the interpretive framework, within which the flow of information and reporting is organised and the dominant views of particular stories created. Some of our interviewees were acutely aware that the media do not 'simply report the world', and equally important that they do not merely 'report the world simply'.

Many of the trade unionists in our sample in fact commented on what they saw as the generally unfavourable descriptions offered of unions by the media and some had individually experienced particularly unhappy clashes with the press and TV.

Some elements of the interpretive framework for the reporting of industrial life are outlined by NALGO in their submission to the

Annan Commission on the Future of Broadcasting. This union argues of the BBC's coverage that:

> The BBC is too establishment-oriented, with its independence ranging within loyalty to a fairly narrow establishment consensus of opinion. This means that its idea of 'balance' in controversial matters and news reporting is too inhibiting and minority opinion is often ignored (e.g. trade unionists are often treated as a minority group – when there are nearly 10 million in Britain!). It also means that assumptions creep into its approach – e.g. the assumption that inflation is wage-related, that any pay increase is a threat to the Social Contract, that any industrial action is reprehensible (with words like 'idle', 'unofficial', 'wildcat' being common). These are common criticisms, but the effect is that the dogged work of trade unionists and their leaders, in securing agreements, in negotiating and in their day-to-day work, is, for the most part, ignored – in spite of the fact that this is a significant feature of a developed, industrial society.[35]

The notion of industrial action as being somehow reprehensible as such, was pointed out on several occasions as a key assumption in media reporting. The National Executive Officer of the AUEW pointed to three main areas in which he believed media bias to be most evident. First, he commented upon interview situations in which questions were loaded in favour of a particular viewpoint; second, he believed that the process of selection and editing was used to underline particular accounts; and third, he pointed to the general assumption of unions as strike-prone in media coverage and to the scapegoating and attacks upon left-wing union members.

First then, he regarded interviews as normally being conducted on the basis of loaded questions. While the issues involved or even who started the dispute, he suggested, are usually ignored, the key question addressed to the union is: 'What are you doing to end this strike?' A supplementary question is 'But is it not true that a large number of your members disagree with the dispute?' The coding of this, he argued, is that first, the unions are responsible for the stoppage and, second, that they are obstructing a settlement. The union official commented that some of the 'strikes' which involved his members, were actually disguised lockouts and had been anticipated, and to some extent planned, by the management – but he suggested that this was not an angle that ever appeared in media coverage. In contrast to the typical questions addressed to the union, the management might be asked 'Is this strike affecting exports?' and 'How much is it disrupting production?'

The union official also suggested that the actual production of, editing and selection of news, often involved distortion. He believed that the general assumptions which underlay media coverage were that strikes were wrong and that workers should not really engage in such things. Union leaders singled out as the causes of such 'disruption' received particularly unfavourable coverage, e.g. the salary of H. Scanlon (but not of, say, the Director-General of the CBI) would be under constant scrutiny. The National Industrial Officer of the MGWU gave a graphic example of this from his own experience. Through a secretarial error a statement on the EEC by this official was withdrawn from one news service tape and then re-released with a piece missing. Thirty-six newspapers carried the story that he had been 'gagged' on the Common Market. A second news service had simply carried the full statement, and one national newspaper actually carried the 'censorship' story on one page and then the full statement on the next. Only one journalist telephoned the Union headquarters to find out if the story was true. The General Secretary of ASTMS took up the issue of the normative assumptions built into media coverage. He argued that the coverage of industrial affairs in Britain was built largely around the presupposition that strikes were in some way sinful or morally offensive. He commented to us that by contrast a major steel strike in America, for example, would be regarded as a simple market phenomenon.

Other interviewees commented upon the contradictions in the picture of industrial life which is given in the media. The Head Research Officer of the TGWU pointed out to us that while much coverage is devoted to the faults of 'unofficial strikers' as 'wildcats', etc., the media are quite capable at the same time of presenting union leaders as authoritarian and as exercising total control – 'the new bosses', etc.

Within the stereotypical presentation of union activity, some interviewees pointed to the use of different stereotypes; for instance middle-class occupational groups such as consultants receive a more favourable coverage when engaging in 'disruptive' activity. In some cases, it appeared that these differences reflected the organisational structure of the broadcasting institutions themselves. The General Secretary of the National Union of Teachers was basically fairly pleased with the coverage that he and his members received, but he commented to us that he believed the BBC's coverage of educational issues was particularly good. Unlike ITN, the BBC sustains an educational correspondent, who is in fairly close contact with the NUT over a wide range of issues. In contrast to this, the General Secretary argued that ITN tend to cover only disputes which will often involve simply sending an industrial reporter. One possible reason for this may be that the BBC perceives itself as a socially responsible institution and thus as having particular obligations towards education.

The Press Officer of NALGO suggested that the groups which it recruited, social workers, etc., might be thought of as the 'backbone of responsible Britain' and commented that the traditional media stereotypes which attached to these groups, were very different from those which attach to car workers or dockers. However, the general increase in white-collar unionism and a growth in the use of militant tactics had seriously confused some sections of the media, the Press Officer claimed. The response of a number of journalists to the 'new NALGO' had been shock and amazement that a 'responsible' union could undertake militant action. In the Press Officer's view, the journalists resolved this by scapegoating individuals, sometimes 'nameless' as the cause of the 'responsible' public employees being led astray. NALGO thus received a spate of coverage in the press under headlines such as 'Reds under the Beds Row Rocks NALGO', 'Revolutionaries – NALGO Warning', 'Fight to Save NALGO from Red Wreckers' (Local news, *Daily Mail* and *Sun*), and the more subdued 'Left-Wing Threat to NALGO' in the *Sunday Times*.

The Assistant General Secretary of NUPE made similar points about the scapegoating of individual trade unionists. Of course the public sector unions are not the only ones to receive the 'red menace' brand of coverage, but why should this interpretation be taken up and used by the media at this particular time? A possible explanation which emerged from discussions with NUPE and NALGO was that the media work within an ideological framework of consensus, an element of which is that if everyone works hard and co-operates then all will prosper. Within this view there is a limited range of explanations for conflict and crisis and such explanations as are used by the media, amount largely to blaming the workforce. Critiques of the economic system as such appear to be almost totally prohibited within the media framework of explanation.

Within this limited range of inferential frames to explain crisis, there are two which feature predominantly in descriptions of the workforce and unions. First, that the conflict is the result of industries being strike prone; and secondly that the strikes, militancy and unrest, are often caused by a small minority.

Both of the interviewees from NUPE and NALGO noted these two kinds of explanations and commented upon them. They both believed their unions to suffer more from the second stereotype than from the first. They independently pointed to the dispute in March 1975 over private beds at the Westminster Hospital and to the scapegoating of the union spokesman by the media as some form of 'loud-mouthed agitator'. The interpretive framework which was adopted in some news bulletins, was that of the union leading the unwilling or unknowing majority.[36]

The Assistant General Secretary of NUPE suggested to us that the media had resolved the intractable fact that the membership were 'responsible public employees' and the difficult social questions posed by the pay beds dispute by first degrading the status of the employees and second by making the principles involved nothing more than the machinations of agitators. He commented that the general framework used in much reportage of the pay beds disputes was that of 'a lot of hospital porters, led by bolshie agitators who think they can tell important people like consultants what to do'.

A further example of the media producers resolving ambiguities via a taken for granted, ideological framework was offered by the Press Officer of the nurses' union, COHSE. He argued that amongst the 'ground rules' of media coverage was prohibition on general attacks on the nursing profession. The militant activity undertaken by nurses in their pay campaign of 1974 thus created some difficulties by portraying COHSE as an ultra-left union leading its members into militancy and by emphasising the differences between COHSE and the more conservative Royal College of Nursing.

The union representatives from the MGWU and the POEU in explaining that coverage in general was inadequate, argued that technical constraints on the production of news exist. These included limitations of space and time such as schedules for copy and production deadlines. The POEU representative went somewhat beyond these technical limits and spoke of the 'moulds' into which journalists have to fit their copy.

A number of other union representatives, including those of the NUR, NUPE, NALGO and the AUEW, commented on private ownership of the media. NALGO quotes the TUC on this in its evidence to the Royal Commission on the Press. 'The fact that eight men control 90 per cent of Britain's newspapers means that the conception of "freedom of expression" has to be highly qualified, especially when such questions as trade union rights or the ownership of wealth, are concerned.'[37]

The union representatives in the sample from the TGWU, NUR, AUEW, COHSE, NALGO and NUPE explicitly suggested that a key premise in much of the organisation of media coverage was the undermining of the status and legitimacy of trade union activity. These points were made cogently by the General Secretary of the NUR, who argued that the ownership and control of the media was a crucial determinate on the nature of coverage. He believed that the trade unions as political organisations were committed against the capitalist system, therefore, he commented, 'what could they expect?' But this had not deterred his union from organising a very advanced publicity department; rather, the General Secretary argued that the size of the problem made the development of such departments essential for trade unions.

The NUR's General Secretary argued that an interventionist approach to the media was necessary primarily 'to change the image of trade unionists as having horns coming out of their heads', and their stereotyping as creators of chaos and disruption. This was made more immediate as a problem for his union because of what he called the instantaneous effect on the public of disputes on the railways and the rapidity with which the media focused upon them. He also said that the mass media industry represented the most efficient means of communication and that it was essential to at least attempt to use this network in order to inform the grass roots union membership correctly on union affairs. The NUR is the only blue-collar union in our sample to adopt this approach.

The white-collar unions are already beginning to use their more developed publicity operations to challenge what they see as the biased frameworks within which they are reported. NALGO and NUPE are consciously moving in this direction. All of the blue-collar unions in the sample were, to varied degrees, critical of the media, and it may well be that others will follow the example of the NUR. If this is so, then the white- and blue-collar unions may well converge in formulating an extensive critique of the mass media institutions. The most probable demands which would be raised in this would be for a form of 'access news', the joint production of programmes and for national ownership of the press and television and independent monitoring or controlling agencies. These proposals were suggested to us by the NUR and NALGO. In addition, the POEU is at the moment pressing for a national television grid on cable links which would allow for two-way transmissions and a more de-centralised production of television programmes. In fact the ACTT (Association of Cinematographic and Television Technicians), the union, with the ABAS (Association of Broadcasting and Allied Staffs), most closely concerned with television and itself responsible for some serious critiques of the media, submitted the following resolution to the 1975 Annual Conference of the TUC, which was passed overwhelmingly.[38]

> *Countering Anti-Trade Union Bias in the Media*
> Recognising the over-simplification and distortion which characterises the manner in which the majority of the media discuss and report economic issues and aware that this over-simplification and distortion frequently expresses itself in savage attacks on the objectives and methods of trade unions, engaged in free collective bargaining, Congress calls on the General Council to instigate the production on a regular and ongoing basis of a counter critique, deliberately written to correct and counteract the distortions of the media and to provide for shop floor trade unionists a straightforward and effective refutation of anti-trade union propaganda.
>
> Congress believes that despite the useful contribution made by the publications of the Labour movement, the TUC must be more effective in answering the attacks of anti-trade unionists in the media.

The effects of union inputs

In assessing to what extent news coverage is affected by the publicity operations of trade unions one conclusion, that certain issues are singled out in reporting, is important. It is, for example, extremely unlikely that the skewed pattern of the British Leyland coverage would result from the publicity operations of the unions involved; especially since in this case they, the AUEW and the TGWU, both have relatively little formal organisation for publicity. In fact as our more detailed analysis of the car industry shows such union representatives, as appeared, were unable seriously to challenge the dominant themes around which the story was organised. We would argue that the publicity inputs, however efficient, are less important than the basic interpretive frameworks used by the newsrooms, although the presuppositions of the news producers may give rise to significant differences between channels, as, for instance, in the attitudes of ITN and BBC towards education.

In terms of the extent of coverage given by both channels, this opinion is borne out by our research. For example, the NUT conference in March 1975 was featured in seven bulletins in three days on both BBC channels, but was not covered at all by ITN. This pattern is confirmed in coverage of other issues such as the teachers' pay claim of April 1975 which was mentioned in one bulletin each on BBC1 and BBC2 but again none on ITN. The elections held by National Union of Students in April 1975 was mentioned in three bulletins on BBC1 and one on ITN, while the NUS grants campaign later on in May was featured in one bulletin each by BBC1 and 2, but received no mention at all by ITN. The General Secretary of the NUT suggested that ITN tended to favour the more 'sensational' items in its education coverage. It is perhaps significant then, that the main other coverage given by ITN in the period studied was of a teachers' strike at Aylesbury and a demonstration in Richmond against proposed education cuts. These both occurred in February 1975 and were not covered by the BBC. This distinction in educational coverage, embodied, as it were, in the educational correspondent, is matched by the differences in industrial coverage between the services noted in Chapter 4 and again attributed to an organisational distinction. The academic debate as to the autonomy of the producers which arises from the observation studies must be seen in this light. Whether the producers are autonomous as Blumler suggests or constrained as Halloran et al. and Elliot[39] propose, in the newsrooms, we would suggest, autonomy becomes a question of internal organisation, at least as far as can be perceived on the screen.

The assessment of the efficiency of trade union publicity is thus very complex. For individual unions it is possible to count the number of times their name is mentioned and statements from them are carried on issues which involve their members. This gives us a crude measure of how much a single union is featured, but tells us nothing of the 'quality' of the actual coverage. A good illustration of this problem is the coverage given to the National Union of

Public Employees in the first three months of 1975. The main disputes involving NUPE and featured in the news in this period were at the Morriston Hospital, Swansea, and at the Westminster Hospital in London. The Morriston dispute was mentioned in 12 bulletins, 4 on BBC and 8 on ITN. In these, the union was actually named on two occasions by each channel and ITN carried three statements from NUPE. On one occasion a statement by the union was 'balanced' by the inclusion of the comments of consultants at the hospital.

In the Westminster dispute statements from the union were included on 6 occasions and, in addition, the union representative appeared in 6 interviews. In all, the union was named 11 times in a total of 21 bulletins. In the case of Westminster, consultants were interviewed on three occasions and the Chairman of the Medical Executive Committee once. In addition, two other reported statements from patients were included. In both of these disputes, in terms of a formal count, NUPE received more coverage than those who spoke against its actions. But we must look beyond the formal count of 'who and what got on' and examine briefly the quality of the coverage. The Westminster Hospital dispute is important, since it was, in the minds of many of the trade unionists we interviewed, something of a *cause célèbre* – a prime example of media bias. The treatment which Jamie Morris, the union spokesman at Westminster, had received at the hands of the press was featured in a BBC2 programme entitled, 'Don't Quote Me' (21 May 1975). One example that the programme pointed to was a full page headline in the *Sun* which read 'Jamie You're a Bastard'. This crudity is, of course, not permitted by the 'quality' stance of the television bulletins. The position of the BBC was to imply the union was leading an unwilling or unknowing majority. This interpretive framework conditioned part of the coverage the union received.

> At Westminster Hospital in London ancillary workers belonging to the National Union of Public Employees are intensifying their campaign over paybeds. They've decided to operate a work-to-rule which would affect all patients and not just the private wards. Union leader, Mr Jamie Morris, said that so far their action hadn't had much effect. Michael Vestey put it to him that in any case, the majority of his members were foreign and didn't understand what the dispute was about.
> JM: 'I'd reject that; we use interpreters at meetings, obviously;

> we do have a high content of immigrant workers. They understand the principle of the strike, they understand why it's got to be and they support it as heartily as anybody else' (BBC1, 17.45, 18 March 1975).

The BBC thus went so far as to suggest that because the majority of the union membership was 'foreign' they would be incapable of understanding the principles involved in the dispute. The BBC returned to this dominant view on the following day:

> Michael Vestey: 'The decision to escalate industrial action was made by the eight-member action committee of porters and domestics. The committee says it represents all the ancillary workers of the hospital; many of them can't speak English and interpreters were used to tell them of the strike decision. Their leader is hospital telephonist, Mr James Morris.' (BBC1, 17.45, 19 March 1975).

In the coverage of this dispute, there was no factual evidence given to support such a view, e.g. that the actions of the committee were at variance with the wishes of the ancillary workers in the hospital. In the absence of evidence the interpretation is given by implication, e.g. 'The committee says it represents . . .', and the references to 'foreign' workers.

A further aspect of the agenda-setting frame used by the BBC, was the taking up of the hospital management's view that the dispute was 'pointless'. In the BBC's report of 17 March 1975 the management's view is given as follows:

> Tyndall: 'The closure of some public beds to allow building work comes at a time when there is plenty of room available at the hospital for both NHS and private patients. Therefore, says the management, the union's demand for unoccupied beds in the private sector, is pointless. . . .' (BBC1, 21.00, 17 March 1975)

This view is then in the same bulletin put to the union spokesman, in the form of: 'Isn't it [the dispute] rather petty?'

A similar formula for interview questions is used by ITN on the same day, who suggested to the union spokesman that 'it's all a bit pointless, isn't it?' (ITN, 22.00, 17 March 1975)

The defence, often used by media professionals, is that questions are formulated for interviewees on the basis of their 'opposition's' case. As long as this is done equally to all interviewees, then it is claimed, some form of balance is achieved. A striking feature of the Westminster coverage, however, is that while the ITN actually did this and put the union's case to the management, e.g., 'It would surely be very simple just to designate ten private patients' beds to National Health use over the next three weeks. Why don't you do that?' (ITN, 22.00, 17 March 1975), the BBC did something quite different. The above extract from the BBC's coverage of 17 March 1975 continues with:

> '. . . [is pointless] and since the union's withdrawal of services to private patients, volunteers are helping with domestic chores.'
> Interviewee (Chairman Medical Executive Committee): I think that other paid employees of the hospital are helping out to run these services.
> BBC interviewer: 'Are any of the consultants doing any chores as it were?'
> Interviewee: 'Well I know at least one of my colleagues has been washing-up earlier on this morning and I don't doubt there will be many others who will also lend a hand – I'm not bad with a Hoover myself. . . .' (BBC1, 21.00, 17 March 1975)

There were other significant differences between BBC and ITN. On the same day as the above coverage, ITN was describing the dispute in the following way: 'it looks as though the real issue is one of principle' (ITN, 17.50, 17 March 1975), and did address some questions to the union on the lines of 'What was the reaction of NUPE members to volunteers doing their work?'

NUPE told us that they are, in general, unhappy about the mass of coverage which they receive in the press and on television. In the case of the Westminster dispute, the activities of their publicity department were unable seriously to affect or modify the picture which was given of them in much of the popular press and by the BBC. The differences which existed between the coverage of ITN and BBC are unlikely to be explained in terms of the publicity output of the union.

Second, the case of the Westminster Hospital dispute should serve as a cautionary tale to those who might evaluate the 'success' of publicity operations merely in terms of the number of times a union is featured. This point was made very forcefully to us by members of the NALGO publicity department who argued that 'all publicity was very definitely not good publicity'.

There is perhaps no union which illustrates this point better than the AUEW. Engineering as such received very little coverage in the period of our research. It will be remembered that the AUEW official interviewed complained that his union and officials within it, had been the subject of a campaign in the media on issues ranging from the President's salary to the union's election procedures. The main focus of this latter coverage was on the influence of the left in the union, the current elections and whether the union system of voting was to be changed to a postal ballot. There were only nine stories in 22 weeks, although for the whole period engineering was top of the DE's dispute table (see Chapter 5). Scanlon's pay rise was mentioned in one bulletin on BBC1 (7 May 1975) as was the AUEW's opposition to the EEC (BBC1, 24 April 1975). By contrast, the AUEW internal elections were reported in 18 bulletins on both channels (8 ITN, 10 BBC).

It need hardly be said that the cause of such an intense coverage of the internal affairs of the AUEW is unlikely to be found in the publicity operations of the union. None of these stories are concerned with the union as a bargainer for the pay and conditions of its members. The national pay deal for engineering workers was negotiated in March 1975. This was referred to in two bulletins each by both channels, yet although the deal involved nineteen unions, no single one of them was actually mentioned or reported by name. The dominant view used in the period of study for reporting pay settlements, was to evaluate them in terms of the Social Contract. This tendency by the newsrooms to set themselves up as arbiters is well illustrated by the coverage of the engineering settlement. The BBC decided firmly that it was 'outside' the Social Contract:

> A pay deal for nearly two million engineering workers, giving increases of over 30 per cent to those earning minimum rates, has been agreed in London. The deal, effective in three stages over the next nine months, will mean an extra £10 a week. But as actual wages are negotiated from plant to plant, the award will probably have an extra-knock-on effect. It's also still to be approved by the executives of the nineteen unions involved. Our industrial correspondent says that both the agreement's size and its timing so soon after another pay deal

contravene the Social Contract wage guidelines. (BBC1, 21.00, 24 March 1975)

The ITN, however, decided on the same evening that the settlement actually posed no immediate threat to the Social Contract.

> A pay agreement has been reached for an increase of nearly 30 per cent in minimum rates for more than two million engineering workers. It increases the minimum skilled rate from £32 to £36 in May, £40 in November and £42 next February. Our industrial correspondent says this poses no immediate threat to the Social Contract because local agreements, in nearly all cases, mean that engineering workers are already earning substantially above the national minimum. But the agreement which still has to be ratified by the various unions, will have some inflationary effect he said, because it acts as a yardstick by which future company and plant deals will be measured. (ITN, 22.00, 24 March 1975)

One other engineering story received a fairly extensive coverage in the period of our sample. This was the announcement of the closure of Imperial Typewriters in January and the subsequent story ran to sixteen bulletins on both channels, and in only two of these was a statement from a union included. In both cases the name of the union was not given. In one other bulletin a TGWU shop steward was interviewed and named as such.

In the rest of the coverage the TGWU does appear to have a fairly low profile. Much of the publicity which it receives focuses on the personal figure of Jack Jones. In the first three months of 1975 he appeared in television news programmes on fifteen occasions (7 ITN, 8 BBC). There is some rationale for this, since Mr Jones has become one of the most influential members of the TUC's General Council and was heavily involved in the development of the Social Contract. This high appearance rate could hardly be explained by his union's publicity operations, since the TGWU has no formal organisation for the promotion of the personalities in its hierarchy. It is of some interest then to compare his appearance rate with that of another union such as ASTMS whose publicity operation is organised very much around the promotion of the personal figure of Clive Jenkins. He, in fact, appeared only once on national television news in the same three-month period. Therefore, what is the effect of a trade union's conscious attempts to achieve publicity?

The news coverage of the railways in the first four months of 1975 provides some evidence. The three main rail unions were negotiating a new pay deal with British Rail. It was also within this period that the new General Secretary of the NUR was appointed, and embarked on a more aggressive and interventionist strategy towards the media. Some measure of the success of this strategy is given if we look briefly at the coverage received by the NUR from February until the end of April 1975.[40] In this period, when all three major rail unions were standing together in the negotiations, the coverage of the rail pay claim ran to 33 bulletins, the NUR was mentioned in 20 of these by name, reported statements from it were included on 9 occasions and the General Secretary was interviewed on 7 occasions. By comparison, ASLEF was mentioned on 12 occasions, 7 reported statements were included and the General Secretary was interviewed on 4 occasions. The white-collar rail union (the Transport Salaried Staffs Association) was not mentioned at all by name, there were no reported statements attributed to it, and no personnel from it were interviewed. British Rail itself was reported on only 4 occasions and its Chairman was interviewed on 3 occasions. There may then be some correlation between the television coverage and the publicity undertaken by the NUR. ASLEF, however, and in particular, its General Secretary Ray Buckton, received relatively extensive coverage without a correspondingly great publicity organisation. Buckton himself explained this to us in terms of his history as the *enfant terrible* of the media in past railway disputes.

Other unions, including COHSE and NALGO, received a fairly low amount of coverage in the period of study. Here again we encounter the problem of sampling from what is, in practice, an infinite and non-ending flow of coverage, for COHSE in fact received a fairly extensive coverage immediately before our research period and NALGO became most heavily involved in the campaign against local government cuts after we had finished recording. An additional problem is raised by unions such as ASTMS and the GMWU who recruit across a wide range of industries and overlap extensively with other unions. It is thus difficult to establish any clear relation between events involving their members and the generalised reports of those events. Notwithstanding this, there are some general conclusions which may be reached on the dimensions of media production which we have discussed here.

We have in fact noted some considerable differences in the level of resources devoted to publicity amongst the unions in our sample.

The orientation of the blue-collar unions in our sample, with the exception of the NUR, was best summarised by the National Executive Officer of the AUEW. He argued that the level of distortion and bias in the media was such that until now unionists in his situation had regarded it as a waste of time to attempt to redress the balance. The white-collar unions, facing the rather different historical problems engendered by their potential for rapid growth, have developed a more extensive and interventionist approach to the media.

When white-collar unions adopt more militant tactics then they are viewed within the interpretive frameworks which were more traditionally applied to blue-collar union activity, and the coverage they receive will be relatively unaffected by their publicity machines, although it may increase the chances of them being named. Against this, the case of the NUR suggests that a concerted effort by individual unions can have some effect at least in the quantity of coverage received.

If, however, as a number of the unionists in our sample argued, the nature of media coverage ultimately turns upon the ownership and control of media institutions, then it is unlikely that individual differences between unions would radically alter the interpretive frameworks governing the reporting. Our own research certainly indicates that, even given the differences noted above, a rigidity and similarity in the interpretive frameworks is present in both television news services. Inputs from individual unions would not, it seems, radically alter this.

7 DOWN TO CASES

'And now, rubbish'[1]

In the confused debate over the news, one of the commonest charges made against the media is that of 'superficiality' in industrial coverage. Dispute reporting, it is said, concentrates on the 'effects' of union action to the neglect of its underlying causes. The result is that in the absence of essential background information the activities of strikers, which might otherwise appear quite rational and reasonable, are presented to the public as sensational. Underlying such complaints is perhaps a fear of 'trial by television', and the trial analogy is not inapt, especially since such media 'trials' tend to ignore the motives of those publicly accused.

The characteristic inferential framework, used by television journalists in reporting disputes, is to utilise limited aspects of a dispute to create a dominant view. A strike has its roots in grievances or demands but it also has human protagonists and visible results and ramifications. All of this, as in the events in Glasgow, is 'news'. Michael Edwards, industrial correspondent of the *Daily Mail*, remarked in a television discussion: '. . . the background is a fairly static thing, it doesn't change, the background is always the same . . . after that the situation is a moving situation and people want to know what's going on.'

The balancing of these elements is an essential professional task of the journalist. This is not just for the sake of impartiality but also to achieve a literate, easily comprehensible and interesting level of reporting.

'The Glasgow rubbish' or the 'Glasgow rubbish strike'[2] was one of the biggest stories featured in the bulletins during the recording period, covered in 102 bulletins over 3 months from 11 January to 14 April 1975.[3] As regards the coverage given, it could be seen as one of the most important television news stories of the first half of the year. The Heavy Goods Vehicle (HGV) licence holders, working for Glasgow Corporation (over half of them drivers of dustcarts), went on strike. The most public result of this action was that the uncollected refuse was eventually partially cleared by the army. This was the first time in some 25 years that troops had been used in an industrial dispute.

The coverage began with the strike decision and continued periodically as new angles were highlighted until the drivers returned to work on 14 April. The dispute was treated by the Glasgow Corporation and the government as a matter of extreme importance. Yet despite the extensive national television coverage of the issues raised, the actual case of the men on strike was neglected.

The weak position of unofficial strikers in the 'hierarchy of access' to the media contributed to the television news definition of the issues raised by the dispute. Thus the framework used concentrated on issues other than the conflict between Glasgow Corporation and their employees. The focus of the coverage became from the outset a 'health hazard'. Whilst it is in no way suggested that a serious problem did not come to exist as a result of many thousands of tons of refuse lying untreated on open dumps, it seems reasonable to question the manner in which this aspect was established as the initial focus of the coverage, even before the dumps had been created.

The threatened health hazard became the dominant theme of all three news services, being established on the very first day of coverage (11 January). Each of the six bulletins that reported the decision to strike on this day (a Saturday, the following Monday being the first day on which the drivers failed to turn out for work), used library film of piles of uncollected refuse from a strike of some four weeks' duration in the autumn of the previous year. The early evening bulletin on BBC1 included a film report from Glasgow in which Bill Hamilton *voiced over* film of men leaving a meeting with some details of the strike decision. Continuing over *library* film of last October's rubbish piles, he reported: 'It was only last October when rubbish piled up in the city streets for four weeks since Glasgow's dustmen [sic] staged their last strike for higher wages.' All other BBC bulletins on this day used only various cuts of the library footage of autumn's refuse, and in each case the newscaster said that 'the decision to stop work from Monday was bringing fears of a repeat of the situation last October when rubbish piled up in the streets causing a health hazard'. This was seven weeks before 4 March, when Glasgow

Corporation themselves announced that a health hazard existed.

ITN bulletins, whilst using only their own library film of the October dispute, were not so explicit in suggesting the hazard. 'The strike means that Glasgow faces another pile-up of rubbish on the pavements as happened for four weeks last autumn' (newscaster *voice-over* film; both bulletins identical). ITN's *News at Ten* on the following day (12 January) repeated the showing of last October's rubbish, as did BBC1's *Nine O'Clock News* on the 13th. In order to maintain the 'health hazard' angle as the dominant view, this latter bulletin was still for the third day juxtapositioning the library film of rubbish with the current film, by then available, of parked dust-carts. Emphasis was also placed on the fact that this was the second strike by Glasgow dustcart drivers within three months.

That this was the second strike by Glasgow's dustcart drivers within three months of their previous strike was emphasised in 13 of the 14 bulletins covering the strike.[4] This fact was still being reported in ITN bulletins on 24 January (Early Evening Bulletin and *News at Ten*). There was no coverage of the strike on either BBC1 or BBC2 during this week.

There were other disputes in Scotland, indeed in Glasgow, some affecting local authority services, and from the second day of coverage of the dustcart strike, reporting often linked them in bulletins; principally all five bulletins covering the dustcart strike on 13 January (3 ITN bulletins, 2 BBC1 bulletins), BBC1's *Nine O'Clock News* on the 12th, BBC1 Early Evening Bulletin on the 15th, and BBC2 *News Review* on 19 January. For example, 'Another strike is due to start in Scotland tomorrow. Yesterday Glasgow dustcart drivers voted to come out. Today, ambulance officers from all over Scotland decided to stop work over a pay dispute' (BBC1, 21.50, 12 January).

Following a lead report of the ambulance controllers' strike, ITN's *News at Ten* added that 'there will be no rubbish collected either in Glasgow tomorrow when 350 dustcart drivers start another pay strike . . .' (12 January). This packaging led Robert Kee to remark the next day 'they are having a bad time with things in Scotland this morning' (ITN, *First Report*, 13 January). Other bulletins during the next few days, used phrases such as these: 'Still in Scotland . . .'; 'In Scotland ambulance controllers and Glasgow dustcart drivers have gone on strike, in both cases over pay'; 'And still in Scotland, strikes by hospital engineers, dustmen [sic] and ambulance controllers continue.'[5]

In fact, as indicated in Chapter 1, eventually the frame was reinforced for many of the industrial events in the UK during that week. 'The week had its share of unrest. Trouble in Glasgow with striking dustmen and ambulance controllers, short time in the car industry, no *Sunday Mirror* or *Sunday People* and a fair amount of general trouble in Fleet Street . . .' (*News Review*, BBC2, Sunday 19 January).

The story lapsed until the Corporation transformed the media's supposed health hazard by actually declaring that one existed nearly two months later. BBC's *Nine O'Clock News* reported the declaration of the health hazard by the Medical Authority on 4 March. The cause of the hazard was reported to be the piles of rubbish accumulated 'during the strike by dustmen' (not dustcart drivers) and further reference was made only to 'the strikers' in this bulletin. When ITN ran the story for the first time since 25 January, they too were confused as to who was actually on strike, reporting, 'In Glasgow where the dustmen's strike, or rather the strike by dustcart drivers . . .' (*First Report*, 10 March).

On 14 March, BBC1 5.45 p.m. news had reported a meeting of 350 dustcart drivers. But BBC1 Lunchtime Bulletin and the *Newsday* bulletin on BBC2 reported 'the dustmen's strike' and referred to a meeting of '550 men' and '550 dustmen' respectively. *Nine O'Clock News* the same day spoke only of a 'dustmen's strike' (twice). On the *Nine O'Clock News* of 18 March their number had become 500, all of them dustcart drivers. When the BBC1 Lunchtime News the next day reported the arrival of the troops to clear up the rubbish, no mention was made of the strike except via a reference to 'the dustmen who aren't involved in the strike'.

The other BBC1 bulletins, BBC2 *Newsday* bulletin and all ITN bulletins, did however report a 'dustcart drivers' strike' and, further, BBC1 Early Evening Bulletin and *Nine O'Clock News* reported the work of the dustmen 'who haven't been involved in the strike'. The strike is reported as being 9 weeks old on the Early Bulletin and 10 weeks old in the *Nine O'Clock News*.

BBC1's Early Evening Bulletin and ITN *News at Ten*, on the 20th, reported the problems facing the troops in clearing the rubbish, without mentioning the strike. Both the ITN evening bulletins on this day reported these same aspects of the story, but the rubbish in question is that 'accumulated during the unofficial strike by dustcart drivers'. On 22 March the settlement of the similar Liverpool dustcart drivers' strike, after an interim pay award, was reported. Here BBC1's *Nine O'Clock News* mentioned the possibility that 'a similar settlement could be found for the Glasgow strike'. Their Early Evening Bulletin had reported the urging of an interim award 'for the dustcart drivers'. As the bulletins covered the developing, 'moving situation', the selection of information was determined by the new events on each day. Reports were increasingly dominated by meetings of the Corporation, with union representatives, with government officials, and the army and then when the troops finally arrived, by rats and rubbish piles. The dispute itself figured less and less in reports and the information, when it *was* given, varied from bulletin to bulletin, from day to day on the same channel.

On Sunday 16 March, when news was dominated by the expected arrival of the army, BBC1's Lunchtime Bulletin did not mention a strike in Glasgow. The Early Evening Bulletin reported both 'striking dustmen' and some '500 dustcart drivers' – this report also appeared on *News Extra*. The main evening news spoke of 'striking dustmen'. *News Review* accounted for the rubbish pile-up by reference to the 'long unofficial strike by dustmen'.

But before the cause of the strike slipped even further from view, it would be as well to recall how much reporting had been devoted to cases of both sides. For example, on the first day all bulletins which reported that the strike was to occur indicated its consequences in terms of inconvenience and hazard by reference to the previous strike. BBC bulletins reported the decision of a mass meeting to strike from the following Monday. ITN did not report this meeting. BBC1 Early Evening news reported that 350 dustcart drivers had voted unanimously; the other BBC bulletins said simply that dustcart drivers, along with other Corporation drivers, 'decided' at 'a mass meeting'. In these bulletins no numbers were mentioned. Both ITN bulletins on this day reported the request to the TGWU for official backing for the strike, whereas only BBC1's Early Evening Bulletin reported this.

But ITN did not mention the involvement of other Corporation drivers. Both ITN bulletins reported 350 dustcart drivers, only the Early Evening Bulletin of BBC1 reported this number. All BBC bulletins reported the pay claim as 'at least another £3.50 a week', while ITN bulletins reported 'an extra £3.35 pence'. The cause of the demand on all BBC bulletins being 'to bring them into line with private haulage drivers' and on ITN 'to reduce the differential between their pay and that of heavy vehicles drivers working for private companies'.

The dustcart drivers' basic demand for parity appears in all six bulletins on 11 January, but is dropped from both the bulletin reports

(BBC1 *Nine O'Clock News*, and ITN *News at Ten*) on the 12th. On the 13th, BBC1 Early Evening Bulletin reported the dispute as one 'over pay', the other bulletins on this day (BBC1 *Nine O'Clock News* and all three ITN news bulletins) reported the parity claim. The BBC bulletins on 15 and 19 January contained no mention of parity. ITN's *News at Ten* on 23 January has it as 'pay strike' and on ITN's Early Evening Bulletin and *News at Ten* on 24 January the strikers want 'another 14 per cent', 'a further 14 per cent' and they are on strike 'for more pay'. The BBC did not run the story on this day.

Taking the coverage as a whole, out of 40 bulletins covering the dispute on BBC1 only 4 contained reports of the parity claim. Some 7 other bulletins contained references to an 'interim offer' or an explanation such as 'over pay', etc. Of a total of 19 reports in BBC2 bulletins, only 2 explicitly mention a parity claim. Three others report the issue as 'over pay' and one other 'over a regrading structure'. Of the 43 ITN bulletins, 8 report the parity issue and 11 others restrict their reporting of the dispute's cause to simple questions of 'pay', etc. Thus the cause of the dispute is not inevitably reported and, in the minority of cases when it is, almost always characterised as being a pay claim with some references to 'parity' and one reference to 'regrading'.

The confusion of nomenclature and the lack of reference to the unofficial nature of the dispute has to be added to this assessment of the television coverage of the story.

The HGV drivers employed by Glasgow Corporation went on strike in the autumn of 1974 in furtherance of a demand for an extra payment to give them parity with the minimum wage earned by HGV licence holders in other industries. The drivers returned to work after four weeks with the issue unsettled but on an understanding on the part of the drivers that the Corporation would be willing to negotiate a local agreement on the parity issue if the national negotiations, still in progress at that time, did not produce a satisfactory settlement. When the HGV drivers went on strike again in 1975, one of their spokesmen appeared on the regional news programmes reading a statement giving the reasons for their action.

> The committee are still firmly of the opinion that they have a genuine grievance. They believe that the Corporation clearly promised to discuss the issue of a suitable payment at local level if national negotiations failed to provide an acceptable solution. The basic wage of HGV drivers with Glasgow Corporation is

> £32.50, the earnings referred to in last night's programme included bonus and at least 10 hours overtime payment. And there are Corporation drivers who only receive minimal bonuses. The lowest rate for a HGV driver in road haulage is £37.00. The committee are conscious of the effect of the strike on the public and would hope that the discussions they are now engaged in can provide the possibility of rapidly clearing the mountains of refuse now lying around the city.

(Dan Duffy. This statement appeared on *Scotland Today*, and *Reporting Scotland*, 11 March. The STV programme edited out the sentence 'And there are corporation drivers . . . minimal bonuses.' STV followed with an interview with Duffy; the BBC programme did not on this day.) The BBC had mentioned this claim on the *Nine O'Clock News* on 13 January: 'The men say their basic pay is £2.50 less than road haulage drivers *and accuse the corporation of going back on a promise to negotiate a local agreement*' (Campbell Barclay, voice over film of parked dustcarts in shed; our italics).

Thus as reported on regional television and on the occasion noted above, the essential question was one of a claimed promise made by the Corporation to the dustcart drivers to make a local HGV parity agreement. The dominant view of the strike did not generally allow this alternative view to be put. At its most dramatic, this can be seen in the fact that during the whole of the strike, not one of the strikers was interviewed on the national news. Only from the day that the strike ended were the drivers allowed to comment on their lost cause.[6] During the course of the strike and the army clearance operation, 10 other people were interviewed in bulletins on the 3 news services, some of them appearing in several bulletins (the interviews were shown in 20 bulletins in all). The interviewees were as follows: Professor Gordon Stewart, Professor of Community Medicine, University of Glasgow (BBC2, *News Extra*, 4 March 1975; ITN *First Report* (live), 10 March 1975; BBC1, Early Evening Bulletin and *Nine O'Clock News* on 20 March 1975). J. Flockhart, Fire Brigade officer (BBC1, *Nine O'Clock News*, 10 March 1975). Councillor Dick Dynes, Glasgow Corporation Labour group (ITN, *First Report*, 11 March 1975). Sir William Gray, Lord Provost of Glasgow (BBC1, *Nine O'Clock News*, 14 March 1975; *Nine O'Clock News*, 15 March 1975). William Ross, Secretary of State for Scotland (BBC1, *Nine O'Clock News*, 14 March 1975; BBC2, *News Extra*, 15 March 1975; and ITN, *News at Ten*, 15 March 1975; BBC1,

Early Evening Bulletin, 15 March 1975). Mr McElhone MP (Labour) (BBC1, Early Evening Bulletin, 15 March 1975). Lt Colonel Campbell (BBC1, *Nine O'Clock News*, 14 March 1975; BBC2, *News Review*, 23 March 1975). Alex Kitson, TGWU official (BBC1, *Nine O'Clock News*, 14 March 1975; BBC1, Early Evening Bulletin, 14 March 1975; ITN, Early Evening Bulletin, 14 March 1975). George McGredie, TGWU official (BBC1, Early Evening Bulletin, 18 March 1975). 2nd Lt Milne (ITN, *News at Ten*, 1 March 1975).

In addition, 4 bulletins on 19 March showed snatches of conversation between pickets and soldiers (BBC1, Early Evening Bulletin and *Nine O'Clock News*; and ITN Early Bulletin and *News at Ten*). BBC2's *News Review* showed this film on 23 March. Also, following the interview with the Lieutenant, ITN *News at Ten* screened vox pop comments of four soldiers (1 April).

Neither of the TGWU officials interviewed could properly be said to speak for the strikers in an unofficial dispute. Their comments in the bulletins concerned are particularly unrevealing.

Alex Kitson of the TGWU appeared first on BBC1 Early Evening Bulletin and later on *Nine O'Clock News*, giving his reactions to the prospect of the arrival of troops:

> 'I would hope personally that there will be a great deal of deep thinking before that decision is taken and in fact it will be my responsibility to contact the powers that be, and that's the Glasgow Corporation, on this question.'
> Reporter: 'So what will happen now then, Mr Kitson?'
> Alex Kitson: 'Well we're polarised at the moment, I mean I've got to admit that, but we've been in this situation before and we'll have to make arrangements to try and get out, I mean we'll start now to see what we can do to get out of it.'

However, the interview with the same official shown in one ITN bulletin the same day (ITN, Early Evening Bulletin, 14 March), gave information on the strikers' claim, and though the questions put to Alex Kitson are not shown in the bulletin, it is rather more meaningfully integrated into a report of the issues of the strike. A film report began:

> The strikers were defiant this morning as they left their mass meeting with only one of them voting against the decision to stay out. They'd been offered bonus rates to clear up the mess

> which had been agreed with Glasgow Corporation, but the central issue of extra pay for men with heavy goods licences remained unresolved and so the men stay out. Glasgow Corporation insist that this is a national not a local issue.
> Alex Kitson: 'What's got to happen now is that the Corporation and ourselves as a union have got to pressurise for this situation to be cleared up at national level or they go back and fulfil the promise that if it wasn't satisfactorily dealt with at national level they'll negotiate locally, and I think they've got a responsibility to the citizens of Glasgow and in the light of the circumstances, they should get off this hobby horse and take into consideration the problems that the citizens of Glasgow are going through.'

The strain between strikers and their union, since the dispute was unofficial, was not an element of the dominant view. The interviews with the union officials were angled within the dominant view towards the arrival of the troops. Thus follow-up questions as to what action might ensue elicited opaque responses because the officials were referencing a framework otherwise unreported.

The selection of interviewees that appeared in the national news reflects those chosen aspects of the story within which the television bulletins organised their coverage. But the fact that the strikers themselves were not interviewed is not the only reason the dispute was not adequately reported. Those involved in the strike were, on occasions, named, and statements paraphrased by the reporter were attributed to them: 'one shop steward told me today', etc. When the decision was made to send in troops, the Corporation's view was balanced by a clear quote from the strikers, but significantly no mention of the claim is made:

> Tonight Mr Archy Hood, the convener of the drivers' shop stewards' committee, threatened to ask for support from every other trade unionist in the country to protest at the move; he said: 'It's a pretty shocking thing when a Labour Government is using troops for strike-breaking in this way. I've never known a Labour Government to act like this; I think they are trying to starve us out.' (ITN, *News at Ten*, 15 March 1975)

Even given the fact that only officials of various kinds were interviewed during the strike, questions as to the nature of the dispute, the strikers' claim and the Corporation's refusal to meet it, could have been asked.

One film report on BBC1 did come near to the cause of the dispute in summarising the situation:

> The strike began almost 13 weeks ago when the drivers walked out demanding parity with the drivers in the private sector. But the Corporation refused to listen. They have, in fact, all along. The men were told that there would be no local pay deals. If they wanted more money then this would have to come at national level. (BBC1, Early Evening Bulletin, 3 April 1975)

But there was no mention of the claimed promise that had been reported so briefly in the *Nine O'Clock News* of 13 January. The Corporation's case is put very clearly, not the men's understanding of it. Despite the library film used in the early coverage, the basis of the autumn return to work was largely ignored. Only on the day that the strike ended was a striker quoted in a way that related the two disputes on the matter of the promise.

> Dan Duffy, one of the strikers' leaders, said that the men had been starved into submission. He said that the Corporation had ratted on an undertaking they gave last October to hold local negotiations, and Mr Duffy warned that the Corporation's behaviour had done no good to Glasgow's future industrial relations. (ITN, *First Report*, 9 April 1975)

However, ITN bulletins later in the day reported only the defeat of the strikers: 'One shop steward convenor said "there's no future in banging our heads against a brick wall" ' and then 'One shop steward said the strike had been no use because they had been up against the government, the corporation and the troops' (ITN, Early Evening Bulletin and *News at Ten*).

News at Ten also included an interview with Dan Duffy, the first with a dustcart driver, saying that the decision to return to work was taken 'in view of the fact that the Corporation were not prepared to honour any agreement and were continuing to welsh on the agreements they had made with us.'

On the same day the BBC reported at lunchtime, 'One of the leaders of the unofficial strike claimed they had been starved into submission.' This forced submission was an angle chosen by the

BBC to put to Dan Duffy in an interview in the Early Evening Bulletin. Duffy said: 'The Corporation have shown that they're not prepared to honour any commitment that they gave to us, they have welshed on us all along.' This interview was also shown on the *Nine O'Clock News*. The lack of reporting of any 'commitment' or 'agreement' during the strike must, we would suggest, render Duffy's statements rather meaningless for the viewer on the day it ended.

The arrival of the troops 3 weeks before, moved the story away from the dustcart drivers' case. Trade-union opposition to the troops was forthcoming with a demonstration organised by the Glasgow Trades Council on 21 March. *First Report*, transmitted at 1.00 p.m., had no film. The newscaster reported,

> And in Glasgow, the use of the troops to clear the rubbish there has brought a protest march by trade unionists from Glasgow and the West of Scotland. About 500 of them marched through the city centre, this morning. The march was organised by Glasgow's Trades Council. And the spokesman said the army was being used to break the dustcart drivers' strike. The march passed off peacefully though it did meet a certain amount of heckling and abuse from passers-by.

By the time of transmission of the Early Evening Bulletin 5.50 p.m., film of the demonstration was available. Six different shots were included in footage of the Glasgow report, five different shots of the marchers passing by, the third shot in the sequence showing several women standing passively on the pavement. The newscaster read the commentary over the film:

> In Glasgow shop stewards representing 50,000 Scottish workers marched to the City Chambers to protest against troops being used to clear away refuse built up by the 10 week old dustcart drivers' strike and Glasgow's 800 firemen have decided on a policy of non-co-operation with the army. They'll only tackle fires in rubbish heaps if there is a danger to life and property and not while the army are there.

No comment was made on the nature of the march, peaceful or otherwise, or on the way it was received by onlookers.

The importance of actuality film is demonstrated by *News at Ten* on this day. In this bulletin the report of the demonstration is organised around a theme of the public's hostile reaction to the marchers.

Reading to camera, the newscaster's introduction to the report emphasised a new interpretation of the demonstration: 'Earlier today shop stewards representing 50,000 Scottish Workers marched to the city centre to protest against the use of the troops, but as Trevor MacDonald reports, they got little sympathy from the public.'

The film sequence over which the reporter read his commentary now included two shots of people on the pavement as the demonstration passed by; the first in which several women shouted comments (4th shot in sequence), and the second showed two children shouting at the edge of the pavement in front of the onlookers (5th shot in sequence). The 6th shot showed an army landrover passing; shots 1, 2, 3 and 7 in the sequence were of the silent marchers.

> MacDonald: 'If the purpose of today's march was to rally support for the strikers, their demonstration was a disaster. At several points on the three mile route to the city centre, the marchers were shouted at, booed and jeered by people whose rat-infested rubbish has been lying in backyards now for 10 weeks. (shouting) Before the march began the shop stewards warned off members of activist groups who have become increasingly visible on the picket lines since the troops moved in.'

The peaceful march reported at 1.00 p.m. which met 'a certain amount of heckling and abuse from passers-by', and the demonstrations that at 5.50 p.m. simply 'marched to the City Chambers' is presented to the viewers of *News at Ten* as 'a disaster' whose participants 'were shouted at, booed and jeered'.

This demonstration was not covered by BBC1. BBC1's Early Bulletin reported the health problem in Liverpool created by uncollected rubbish arising out of a similar dispute between 300 dust-cart drivers and the Corporation there. Shown in interview on film, the Medical Officer of Health warned of the possible dangers of disease. ITN's 3 bulletins covered the Liverpool situation as well in *First Report*, *News at Ten* (immediately preceding the Glasgow report) and following a Glasgow report in the Early Evening Bulletin. BBC1 and 2 bulletins did not report any of the events surrounding the dispute on 21 March 1975. BBC2's *News Review* however did make reference to the Trades Council march at the end of a report of the Glasgow situation: 'The strikers have made several calls for support in their stand against the use of the army but with little result. One demonstration met with bitter hostility from housewives.' This was read over film of soldiers working at a rubbish tip.

When the drivers began work again on 14 April ITN bulletins reported only that the 350 men were back at work 'after their abortive 13 week strike'. One of the drivers was interviewed and said, in the reporter's words, that he would 'happily go through it all again':

Tom Docherty:	I would go on strike for the same cause again because we're qualified drivers. . . . We're experienced and we're professionals . . . we are entitled to this money. . . .
Rep.:	But going on strike doesn't appear to have achieved anything because the army can come in and do your work?
Docherty:	Certainly, because we didn't have union backing this time.
Rep.:	So why go on strike again?
Docherty:	We'll go on strike on principal and we're still entitled to this money. And there's nobody saying we won't go on strike, we definitely will, if it comes to the cause again, and it's a justful cause, we must go on strike for it again.

The public hardly had a chance to discover from the bulletins what this 'justful cause' was.

'Manifestly avoidable stoppages'

There was 'heavy' and sustained coverage of a strike which occurred at the Cowley plant of British Leyland Motor Corporation by 250 engine tuners, between 3 January and 3 February. This coverage began with the reporting of a speech by the then Prime Minister, Harold Wilson, which he made at his constituency of Huyton on 3 January. This speech covered many areas, notably government policy on industry and investment, but the section of it which received most attention from the television news was a reference made by Mr Wilson to 'manifestly avoidable stoppages of production' in the car industry. This reference was interpreted by all three channels in such a way that they presented strikes as the main problem facing the car industry in general and British Leyland Motor Corporation in particular. Whatever the status and intention of Mr Wilson's remarks, we will show how the media focused on only one interpretation of Leyland's problems – that strikes were their root cause.

The BBC, in its initial reporting of the speech, defined it as being critical of both management and unions. In the first bulletin that evening on BBC1, the speech was introduced as follows:

> The Prime Minister, in a major speech tonight on the economy, appealed to *management and unions* in the car industry to cut down on what he called 'manifestly avoidable stoppages'. He said this was especially important now that government money was involved. The decision to help British Leyland was part of the government's fight against unemployment, but the help couldn't be justified if it led to continuing losses. Mr Wilson singled out for particular blame British Leyland's Austin Morris division, which he said was responsible last year for a fifth of the stoppages in man days lost of the whole car industry. (BBC1, *Nine O'Clock News*, 3 January 1975; our italics)

The bulletin then included extracts from the speech in which Mr Wilson refers to the problems of private capital:

> This is an industry which itself makes a disproportionate contribution to the loss of output through disputes, because with just over 2 per cent of the total employees, 2 per cent of all those working in the whole of Britain, it accounted for one-eighth of all the man days lost in 1974 through disputes and that was a year of course which was inflated by the coal mining dispute, which we rapidly brought to an end, and it accounted for getting on for one-third of the total national loss through disputes in 1973. *Whether this loss of production was acceptable or not with private capital involved, or whether it was simply that private capital was unable to deal with such problems, is a matter now for historical argument.* What is not a matter for argument for the future, is this: With public capital and an appropriate degree of public ownership and control involved the government could not justify to Parliament or to the taxpayer the subsidising of large factories involving thousands of jobs, factories which could pay their way but are failing to do so because of manifestly avoidable stoppages of production. (BBC1, *Nine O'Clock News*, 3 January; our italics)

In the next bulletin, $1\frac{3}{4}$ hours later, two significant changes were made. First, the speech is introduced as applying only to workers, and second, the reference to 'private capital' is missing.

The later bulletin starts as follows:

> The Prime Minister has appealed *to workers* in the car industry to cut down on avoidable stoppages. He said the industry had a record of strikes out of proportion to its size, and he singled out, for particular blame, British Leyland's Austin Morris division, which he said was responsible last year for a fifth of the industry's lost production through strikes. Mr Wilson said that unless labour relations improved, government help for British Leyland would be put in doubt. (BBC1, 3 January; our italics)

The bulletin then showed the following extracts from the speech:

> Parts of the British Leyland undertaking are profitable, others are not, but public investment and participation cannot be justified on the basis of continued avoidable loss-making. Our intervention cannot be based on a policy of turning a private liability into a public liability. [*BBC cut here.*] What is not a matter for argument for the future is this. With public capital and an appropriate degree of public ownership and control involved, the government could not justify to Parliament, or to the taxpayer, the subsidising of large factories which could pay their way, but are failing to do so because of manifestly avoidable stoppages of production. (BBC1, 3 January)

The speech now is defined as being directed solely at workers and the reference in his speech to the problems of private capital has been cut.

The BBC2 coverage on that evening interpreted the speech as being critical of both management and workers. It introduced the speech as 'a blunt warning to the car industry':

> The Prime Minister tonight gave a blunt warning to the *car industry* that public money would not be used to subsidise firms who were losing money because of strikes. Mr Wilson speaking to his Huyton constituency said the decision to help British Leyland was part of the government's fight against unemploy-

> ment. The aim he said, was to put production, exports and jobs on a secure and profitable basis, but the success of public intervention meant a fair day's work for a fair day's pay by *everyone* involved. (BBC2, 3 January; our italics)

In the same bulletin, there is an interview between the newscaster and the BBC's Midland industrial correspondent, in which criticism of management is explicitly referred to:

> Newscaster: 'Many of the phrases in the Prime Minister's speech are pointed directly at the unions and the labour force, some are pointed at management, like the need for more efficient working methods. Do the management accept that they have got to do some pretty radical rethinking about production methods and that sort of thing?'

ITN's coverage embraced from the outset the view that the speech applied only to the workforce. Its main news began:

> The Prime Minister tonight defended the government policy of stepping in to help companies where jobs were threatened but he also gave *workers* a blunt warning. He said the government couldn't intervene just to turn a private loss into a public loss, it was up to *workers* to cut what he called 'manifestly avoidable stoppages' and to curb unrealistic demands. (ITN, *News at Ten*, 3 January; our italics)

Using a similar formula this introduction was then followed with extracts taken from Mr Wilson's speech, but this time they are accompanied by summaries from an ITN reporter at Huyton. Later in the same bulletin the reporter gave his view of what Mr Wilson had said:

> This was a stern message to come from a Labour Prime Minister, but it was received politely enough by the audience here in a Labour Club in his constituency; but the speech was clearly prompted by the growing number of companies going to the government for help and the large sums of public money involved. Mr Wilson clearly expects a greater degree of restraint from the *work force* in firms where the government has stepped in to help and he has appealed directly to *working*

people not to rock an already very leaky boat. Glyn Mathias, *News at Ten*, in the Huyton Constituency. (ITN, *News at Ten*, 3 January; our italics)

So the first part of this bulletin is taken up with the reporting by ITN of those parts of the speech which they consider most newsworthy – that is, what they termed 'Mr Wilson's blunt warning to workers'. The focus from the outset, then, is on the workers and on strikes, rocking 'an already very leaky boat'. The Prime Minister's speech was to be referred to many times in the future – but the BBC2 reference at 9.15 was the last time that television journalists reported it as being in any way critical of management. The dominant view of British Leyland's problem was thus established.

There were alternative accounts of BLMC's problems. Many informed sources were already pointing to a history of bad management and the lack of effective investment. The journal *Management Today* in August 1972, more than two years before this coverage, had made the following points about Leyland: 'Capital expenditure had been very low for many years, and depreciation was correspondingly small. The high profits about which so many boasts were made, were thus derived from a declining asset base and too high a proportion was paid out to shareholders.' The government's Ryder Report on British Leyland, which appeared in April of last year, confirmed these criticisms.

The Ryder Report found that one key problem underlying Leyland's collapse was the lack of effective investment by Leyland's management. In particular the report criticised the management's decision to distribute 95 per cent of the firm's profits between 1968 and 1972 as dividends to shareholders and the retention of only 5 per cent for re-investment purposes. Leyland's profits between 1968 and 1972 had been £74m – of this £70m was distributed as dividends and only £4 was retained for re-investment.

Throughout the period of this case study, the *Financial Times* was carrying reports on the extent of managerial incompetence at British Leyland. One such report read: 'Cowley shop stewards tell hair-raising stories about managerial failings, and point at the moment to constant assembly-track holdups caused by nonavailability of supplier component parts' (*Financial Times*, 6 January 1975).

However, the dominant view as presented to the viewer of the television news was organised almost exclusively around the theme of strikes and disruption as the source of Leyland's problems. The alternative accounts did receive brief mention. The ITN bulletin on 3 January illustrates how this was done. The item consisted of 4 minutes 50 seconds on Mr Wilson's speech and Glyn Mathias's summary. There then followed:

> Mr Wilson's comments on British Leyland got a cool reception from one Labour MP, Mr Leslie Huckfield of Nuneaton. He said the Prime Minister clearly knew very little about the car industry, the real cause of the trouble was the chronic failure of management to invest, he said. But the opposition's employment spokesman Mr James Prior, and the British Leyland spokesman, both supported Mr Wilson's remarks. Mr Prior said Mr Wilson was at least stating some home truths which others have been expressing for a long while. (ITN, *News at Ten*, 3 January 1975)

Mr Huckfield got exactly 15 seconds and his view was immediately countered by the two other quotations. The bulletin quoted above goes on:

> '*As if to underline Mr Wilson's remarks*, British Leyland's Austin Morris plant in Cowley announced that 12,000 men are being laid off because of a strike by 250 workers. The striking workers are engine tuners, who want to be graded as skilled workers. They rejected a plea to call off the strike which could cut production by a thousand cars a day. (ITN, *News at Ten*, 3 January 1975; our italics)

The BBC1 bulletins quoted above, used a very similar format for the presentation of Mr Huckfield's comments. First they are 'sandwiched', but instead of the 'as if to underline' formulation the BBC uses the phrase 'and there was more trouble today'.

> Mr Wilson's speech has been welcomed by the opposition spokesman on employment, Mr James Prior. He said the Prime Minister had told car workers some home truths, although it was a pity he hadn't done so before, but Mr Leslie Huckfield, a Labour MP with a lot of car workers in his constituency of Nuneaton, said the speech was disgraceful. The real culprits were the management, not the workers. British Leyland said tonight they shared Mr Wilson's exasperation at the series of futile disputes in the corporation *and there was more trouble today*. 12,000 workers at the Cowley plant near Oxford were laid

> off because of a strike by 250 in the tuning department. (BBC1, *Nine O'Clock News*, 3 January 1975; our italics)

The Wilson speech becomes defined as relating only to Leyland's workforce. This definition is then used to underline a particular explanation of the problems of Leyland in the reporting of the strike at Cowley. In subsequent days the speech was frequently recalled. It was transformed from 'manifestly avoidable stoppages' to 'Mr Wilson's speech on avoidable strikes' (5 January, 18.55, BBC2), or 'senseless strikes in the car industry' (BBC2, *News Extra*, 9 January 1975).

In all, 7 times on BBC1, 6 times on BBC2 and 16 times on ITN, the speech was referred to as relating to the workforce and strikes exclusively. While the comments of Mr Leslie Huckfield simply disappeared after the second day of the coverage, those of Mr Wilson were used again and again to point to a dominant media interpretation of the strike at Cowley. The typical pattern for this news reporting was as follows:

> Cowley in Oxford, *specially picked out by Mr Wilson in his warning last night about strikes*, is at a standstill for a second day because of industrial trouble. 12,000 workers at the plant are being laid off because 250 engine tuners who want to be higher graded are stopping work on Monday. In his speech last night, Mr Wilson warned workers in general, but car workers in particular, that the government could not justify subsidising large factories which were losing money because of manifestly avoidable strikes. The speech has been welcomed by some Conservative MPs, but condemned by some left-wing Labour members. (ITN, Lunchtime, 4 January; our italics)

> First the fresh strike at British Leyland's. The management at Cowley said this evening that despite the renewed stoppage by the 250 tuners there, they have managed to achieve 80 per cent of a normal day's output. The 12,000 other people who work at Cowley, *the plant which was specifically mentioned by the Prime Minister last week when he talked about senseless strikes in the motor industry* – they were angry this morning when they learned that the tuners had voted to walk out again and that they faced the threat of layoffs for the second time in four days. (BBC2, Late Night Bulletin, 9 January; our italics)

> Meanwhile British Leyland, and this morning the 250 engine tuners at the British Leyland Cowley plant voted by a narrow majority to resume their strike over their claim to be reclassified as skilled workers, had gone back to work for two days while new talks with the management went on. But these talks, the engine tuners say, were unsatisfactory, got them nowhere and now they're out again, *within a week in fact of the Prime Minister's warning that what he called unnecessary strikes were putting jobs in the car industry at risk*. And indeed as a result of this action this morning, 12,000 other British Leyland car workers may well have to be laid off immediately. (ITN, *First Report*, 9 January 1975; our italics)

> The men at Cowley had begun an indefinite layoff period only this morning after 250 engine tuners went on strike. But after talks between the management and union officials at the plant, *which was singled out for mention by the Prime Minister in his warning about strike-prone industries last week*, the company decided to call them back from tomorrow. And engine tuners' shop stewards were meeting their union officials late this afternoon and it is hoped proposals for them to return may be put to the men in the morning. (BBC1, Early Evening Bulletin, 6 January; our italics)[7]

In the above forms and in other forms such as interview questions, this dominant view of the Cowley strike taken from the Wilson speech, was referred to a total of 13 times on BBC1, 8 on BBC2 and 21 times on ITN.

In total then this amounts to 42 references (to strike and speech) emphasising this interpretation. This may be compared with the reporting of a statement made by Jack Jones criticising Leyland's management two days after Mr Wilson's speech. This only received three references on BBC1, none on BBC2 and 3 on ITN. Significantly, the Jones statement was not subsequently used to organise coverage, unlike the Prime Minister's speech, which was still being referred to 17 days after the speech was actually made. The only reference made to Jones's statement subsequently was as part of a question, put to the Leyland industrial relations manager, by Robert Kee. It was one of the 7 questions put to the Leyland Manager:

1 Would workers at Leyland approve of a one-year strike truce?

2 Jones yesterday said management was largely to blame for stoppages – how do you take this?
3 What are stoppages caused by?
4 Isn't this a criticism of the unions as they apparently have so little control over their men?
5 Do you think there are people at work at Leyland who simply want to disrupt the thing?
6 What danger is there to jobs in Leyland if these sort of strikes go on?
7 With government coming to the aid of British Leyland, aren't workers going to think their jobs are safe anyway?

Interviews are particularly illuminating examples of how agenda-setting is used to maintain the dominant view. When the strikers, for instance, attempted to give an alternative version of the causes of Leyland's economic problems, the television news returned persistently to the theme of irresponsible disruption. In the interview quoted above (p. 224) with Doug Hobbs, the shop steward convenor at Cowley, alternative information that seriously questioned the framework of the reporter was provided:

Doug Hobbs: . . . Since April of last year, we have worked consistently, all of us, to try and avoid any disputes whatsoever. In fact most of the production that has been lost, has been lost through either breakdowns or shortage of materials and we do recognise that British Leyland is, has got a problem, a cash-flow problem and we have worked very, very hard, both union and members, to try and eradicate this position.

Harris: How does the prospect of no government cash for British Leyland strike you if the strike record doesn't improve?

Doug Hobbs pointed out that the majority of stoppages at Cowley did not result from industrial action at the plant. This view of Leyland, that its problems were also caused by supply difficulties and breakdown of plant, was in conflict with the dominant view thus far presented by the television.

In fact, figures published five days after this interview in the *Guardian*, *Daily Telegraph*, *The Times* and *Financial Times*, showed that in the immediately foregoing period over half of the loss of production at Leyland had nothing whatever to do with disputes at British Leyland. Thus in the *Daily Telegraph*, 'the causes for the loss of production were about equally divided between disputes and matters the management must put right' (*Daily Telegraph*, 11 January 1975). These figures were never quoted in any bulletins. This alternative view was being offered to one small section of the news-consuming public, that is the readership of the quality press. Other papers took the dominant 'workshy labour force' line. But it is no part of the function of television to duplicate the partisanship of the press. Indeed it is expressly forbidden to do so.

Thus the Hobbs interview becomes an important occasion since it is, at this stage of the story, the only time the television newsrooms mention these other elements in the Leyland story. It is therefore noteworthy that this information given by Hobbs had no discernible effect on the interviewer, who persists with the assumptions with which he begun. Immediately after Hobbs gives the news about breakdowns, shortage of materials and a cash-flow problem, Harris continues with a question about the effect of the strike record on the prospects for public aid. This phenomenon, which we shall call persistent agenda-setting, is dealt with in Volume 2.

The interviewer persisted with the established agenda in the face of the interviewee twice indicating that the incidence of strikes at Cowley had actually declined and that production losses arose largely from factors over which the workforce had no control.

Could it be that the shortness of the bulletins and the time allowed to prepare them did not permit the alternative view to be reported? In view of the great amount of time devoted to the car industry during this period this seems, of itself, unlikely. Cowley was reported 94 times during January and February. The car industry as a whole received 25 per cent of BBC1's industrial reporting, 24·9 per cent of ITN's and 21·8 per cent of BBC2's.[8]

The dominant theme of the television news was overwhelmingly that strikes and the workforce were the source of Leyland's problems. This assumption was revealed very clearly by Robert Kee on ITN:

> Mr Sanderson's call follows the Prime Minister's statement of course at the weekend, when he criticised British Leyland or rather its workers for, well, British Leyland, for its poor strike record and giving a blunt warning that firms with a history of what Mr Wilson called 'manifestly avoidable stoppages' could not expect to be baled out forever with public funds. (ITN, *First Report*, 6 January 1975)

The pervasiveness of the strike theme as an explanation of Leyland's problems may be illustrated numerically by comparing references to the problems that strikes were causing Leyland against references to managerial problems and investment. It should be pointed out that these figures do not include references to the Prime Minister's speech.

On BBC1 there were 22 references to the strike problem of Leyland, as against 5 references to the problems of management and only 1 to investment. On BBC2 there were 8 references to the strike theme, 3 to management and 2 to investment. On ITN there were 33 to the strike theme, 8 to management and none to investment. Since information which contradicts the dominant view tends to be discounted, sandwiched or overwhelmed, these figures overstate the actual emphasis given to alternative explanations.

It can be further noted that the dominant view of the car industry is used to interpret stories in the area which are not primarily about disputes. For example, the industry as a whole was suffering, as part of a world-wide phenomenon, falling sales:

> On the day that it has been announced by the Government that new car sales last year were down by 25 per cent on 1973, the Director of British Leyland's Cowley plant has warned of a calamity if the strike situation there gets worse. Figures out today show that private car and van registrations dropped from 1,688,000 in 1973 to 1,273,000 last year and all vehicle registrations were down nearly as much by 20 per cent. The warning came in a letter from the Plant Director, Mr John Symons, to Leyland employees as the Company and the Engineering Union agreed to talks tomorrow at the Conciliation and Arbitration Service to try to solve the strike of engine tuners at Cowley. Mr Symons said that the strike had meant that Cowley was failing to meet what he called its survival budget. He also gave a warning that a further deterioration would be calamitous with the strongest likelihood of a major reduction in manufacturing and employment at Cowley. (ITN, *News at Ten*, 22 January 1975)

Were then cars not being bought because they were not being made?

> Twelve months ago more than 5,000 people bought a Chrysler

> Avenger, now the latest monthly figures show only 3,700 sold. With a touch of irony, Chrysler point out that they'd had a run without production stoppages without strikes, so stocks must be good, but for all their optimism their workforce is going on a three day week for the rest of January. (BBC1, *Nine O'Clock News*, 9 January 1975)

Perhaps the most remarkable effect of the dominance of the strike theme as an explanation is that on occasion, television journalists can actually use this sort of factual information to bolster the dominant view even further. This is illustrated most dramatically with the coverage of the Ryder Report on 24 April, whose findings contradicted the pattern of explanation we have just outlined. Despite the Ryder Report's rejection of the view that strikes were the sole or indeed the major cause of British Leyland's problems, the ITN main bulletin that same day actually reversed the sense of the report:

> Ryder says his team does not subscribe to the view that all the ills of British Leyland can be laid at the door of a workshy labour-force, and the Prime Minister emphasised in the Commons that unless there were fewer stoppages and higher productivity the government would not feel obliged to keep putting money in. (ITN, *News at Ten*, 24 April 1975)

Not, 'however the Prime Minister, etc.', or 'but the Prime Minister etc.'; just simply 'and the Prime Minister . . .'. And it should also be remembered that the government Arbitration and Conciliation Service vindicated the tuners' basic claim for regrading when eventually they reported.

Our analysis goes beyond saying merely that the television news 'favour' certain individuals and institutions by giving them more time and status. Such criticisms are crude. The nature of our analysis is deeper than this: in the end it relates to the picture of society in general and industrial society in particular, that television news constructs. This at its most damaging includes, as in these case studies, the laying of blame for society's industrial and economic problems at the door of the workforce. This is done in the face of contradictory evidence which, when it appears, is either ignored, smothered, or at worst, is treated as if it supports the inferential frameworks utilised by the producers of news. It is by these strategies and techniques, encased in the overall structure of the news bulletins, that we have endeavoured in this volume to reveal, that the ideology of 'neutral' news achieves its credibility on the screen.

APPENDIX I

Four of the categories were divided into sub-categories. Six were not subdivided. The lighter categories (in press terms) of sport, human interest and disasters, remained as wholes. The only problem was in the disaster area. It was decided to categorise near misses as disasters (e.g. aircraft mishaps), all deaths by accident as disasters even when only one person was involved. But the deaths of the famous, that is where obituary material was broadcast, were categorised as human interest. This area was defined quite strictly, and we sought to place stories perhaps conceived of as human interest into other more concrete categories, e.g. collapsing houses into disaster or home affairs. But we did include royalty in this category; this caused some problems during the second phase since a major political story blew up about the Queen's finances (10) at exactly the moment she was on a state visit to Mexico (80).

Science was largely determined by the presence of the news services' science correspondent. It was not subdivided. It overrode the foreign category since space stories, etc., are normally the province of the science correspondent and not the foreign correspondent in the country concerned. This is not, however, true of crime and, therefore, the category crime did not override the category foreign. That is to say, crime stories from abroad were categorised as foreign stories.

The industrial category was not broken up because we knew that a major part of our work would be involved in the detailed analysis of industrial stories and that we would have them to hand on the archive video-cassettes for further breakdown.

Thus politics (10), foreign (30), economics (40), and home affairs (60) were all subdivided.

10—11 Parliamentary politics
12 Northern Ireland
13 Terrorist
14 Demonstrations

15 Labour Party internal
16 Conservative Party internal
17 Liberal Party internal
18 Other parties and pressure groups
19 Politicians personal

We were less interested in Westminster stories than the newsrooms are. Therefore we tended to reallocate Westminster stories by their subject matter – so that, for instance, government monthly figures would be placed in the appropriate economic category, the doings of the Secretary of State for Northern Ireland in the Northern Ireland category and so on. We included in Northern Ireland (12) all activities of the IRA in the Republic of Ireland as well as in Britain, all Republican political news if it related to the North. The main problems here were that bomb explosions in the United Kingdom outside the Province were not always attributable to, or indeed attributed to, Irish activists. Therefore, they would be placed in category 13 (terrorist). Rarely would a bomb story remain without a suggested source for the device for more than one bulletin. If it was then attributed to Irish groups all references were placed in 12. Irish Republic stories not relating to the north were categorised as foreign (30). The trials of bombers were categorised as 12 not as crime (50) and this applied also to the trials of other non-Irish 'terrorists'; they were 13 not crime (50).

During this period demonstrations were not generally reported as events in themselves but rather related to the demonstrators' grievances. Thus, protests against imports were placed in the general industrial category (20).

Category 18 relates not only to parties represented in Parliament (SNP, Plaid Cymru) but also to those unrepresented (Communist Party of Great Britain, etc.) and to pressure groups whose activities are directed to Parliament. (These would include the various groups established around the Referendum campaign over Britain's continued membership of the EEC.)

Category 19 relates to those stories which would otherwise be 80, human interest, if the principle figure was not a politician. Mr Heath's yachting was the example in mind. As it happened nearly all stories subsequently categorised as 19 related to John Stonehouse and that, therefore, in this connection, 19 overrode 50, crime.

Categories 15, 16 and 17 are self-explanatory.

30—31 Britain and the World
32 The World

It was thought that this area would be broken down geographically and categories for each area of the world were determined. But after the initial dry-runs we decided to simplify these inputs mainly because it was not an area of primary interest to the project. Only the balance between domestic and foreign was deemed important. Britain and the World (31), included UK foreign policy, all EEC stories, except the Referendum campaign which was nearly always reported as a domestic political story; Britain and the UN and Britons involved in events abroad. 32 included all foreign stories except sport, disasters and science. Thus industrial and crime stories from abroad were classified as foreign. Some economic stories were classified as foreign as well – for instance, reports of other countries' balance of payments, unemployment, etc., were foreign but international commodity and currency markets were classified as economic. Foreign-industrial and where appropriate, foreign-economic were kept and transferred to the archive.

40—41 Economics, City
42 Economics, Currency
43 Economics, Commodities
44 Economics, Business

The city category included stock market stories, especially the regular reporting of such indicators as the Financial Times Share Index (FT Index) as well as the government's balance of payments statistics. Category 42 included the performance of the Pound and all stories relating to the international currency market. It overrode the category foreign since it was normally reported by the economics correspondents. 43, commodities, included world fuel stories as well as the operation of the international food markets. It did not include gold, which was categorised as 42, nor did it include domestic consumer stories about food prices which were categorised under home affairs. This category too overrode the foreign unless the story was presented as an entirely domestic problem for the country concerned – i.e. third-world food riots would be 32, but such riots related to the operation of the Chicago Grain Market would be 43. Category 44 was concerned with the performance of business, especially as it turned out, with business collapses. It thus overrode category 41 when company shares were suspended; and overrode 50, crime, when companies were accused of fraud. But if such stories included an element of short-time working, layoffs or dismissals they would be categorised as 20, industrial. Pronouncements by the CBI on business performance as well as Parliament's intervention in the affairs of business, would be 44. The general tendency however, was to move such stories towards the industrial category if labour was at all involved in the reporting of the situation.

60—61	Home Affairs, local government	63	Home Affairs, the quality of life
62	Home Affairs, consumer	64	Home Affairs, race

The local government category included all stories relating to charges and services made and provided by local government including primary and secondary education. It also included local government corruption and thus overrode crime (50). Category 62 included public service utilities as well as consumer stories relating to goods and services. The environment (63) included pollution stories (but not when they occurred abroad), health (but not in its industrial or scientific aspects), and all stories relating to the preservation of the environment in town and country. Race (64) included all immigration stories as well as reports from the countries from whence immigrants came. It did not include stories of Britons as immigrants which were normally classified as 31, Britain and the World.

APPENDIX 2

Along the top of the sheet is the basic identification information. Also across the top are spaces for the names of the television personnel involved in the broadcast and some aide-mémoires to subsidiary classification systems which will be explained below. All information as to date, channel, time and logger (i.e. the person responsible for the log), were standardised to expedite the transference of the information onto computer coding sheets. Thus the date became 263 rather than 26 March 1975 or other variations and the project members were assigned numbers to be used in the logger box.

The time of transmission was normally the advertised time according to the programme journals and the lateness of bulletins relative to such times was not noted. Anyway the BBC1 main bulletins at nine and ITN's *News at Ten* represented fixed scheduling points so that it was only BBC2's *News Extra* during phase two of the project that was regularly broadcast minutes after its advertised time.

The first column on the sheet below these basic identification boxes is for the item number. We did not note the opening titles of bulletins when they did not involve the use of different newsfilm or photographs which changed from bulletin to bulletin. Sometimes, however, a major story would cause a change in the opening which would be noted; e.g. the substitution of Mrs Thatcher for Mr Heath in the standard *First Report* opening title after her election as leader. With the major bulletins, BBC1 at 9 o'clock and ITN's *News at Ten*, the bulletin began with headlines unillustrated on the BBC but illustrated on ITN and punctuated there by a single chime of Big Ben. In ITN these headlines are known as 'the Bongs'. We noted the contents of the headlines but did not number them as items. Similarly at the end of the first part of *News at Ten* the newscaster reveals the major stories to be seen after the commercial break. Here again we did not number this *hooker* but noted its contents. Some bulletins close

DATE	1-3		PERSONNEL CODE		COR.B		REP. 11		REP.
CHANNEL	5		N/C 1		COR.C		REP. 12		REP.
T/X	6		N/C 2		COR.D		REP. 13		REP.
LOGGER	7		COR. A		REP. 10		REP. 14		REP.

No.	HEADLINE	SUBHEAD	Overall DUR	PERSONNEL (see Code)			INTERVIEWEES	
				IN	V/O	INT.	NAME	STA
10-11	12-13	14	16-18	20-21	22-23	24-25		

Appendix 2, Figure 1 Initial logging sheet

INTERVIEWEES:	*GRAPHICS:*	*COMMENTS, e.g.:*
1. Central Fig 2. Spokesman 3. Opp. Spokesman 4. Vox Pop 5. Witness 6. Conf. 7. Expert	1. Cap Sup 2. Pix 3. Graphic 4. Anim. Gr. 5. Map 6. Wire 7. Cap. Seq. 8. R.C. Cap.	Non Broadcast Film Source Amateur Film Source Promotions Public Service Women Blacks Length of Item Multiple Inputs

TECHNICS																T/F	ARCHIVE	COMMENTS
FILM				Live	VTR							Lib	RC	GR see code				
SOF	F/X	BW	Pool			St.	REM	OB	EBU	Sat.	Pool							
27	28	29	30	31	32	33	34	35	36	37	38	39	40	41-48		49		

© COPYRIGHT S.S.R.C. MEDIA PROJECT, GLASGOW

with a straightforward and unchanging formula. The main BBC1 news repeated the main points at the end of the bulletin. *First Report* on ITV always concluded with a short item giving the Financial Times Share Index and weather for the day. We did not number these items. Therefore, only the changing items in the bulletin proper excluding openings and closings were numbered.

The tendency of British newsmen to package has been noted (F. Rositi et al., *News and Current Events on TV – An Analysis of Television News on Four European Television Organizations*, Prix Italia Report, Florence, 1975), and above we commented on the difficulties this raised in trying to itemise the bulletins. Indeed the lengthier the packages the better the news is seen to be in the eyes of those producing it (see Chapter 3).

Apart from personnel changes noted above, the guidelines to whether one was dealing with one item, two items, or one item segued out of another, were in the scripts themselves. Thus, a sequence of stories concerned with the parliamentary row about the Queen's finances could be followed by the Queen herself in Mexico and Prince Charles on his way to Katmandu for the royal wedding. This could be one item if introduced as a Royals package or two items with the royal journeys seguing out of the political story and the Charles element being a payoff to the Mexican story, depending on the form of the script. Statistical stories and general city stories are often used to introduce or payoff and close particular business and economic stories. In such cases these elements would count as one item with the particular story so introduced or paid-off.

On occasions there are breakdowns in the presentation and items are abandoned to be picked up later in the bulletin. In such cases the initial item number was re-used. In stories which were breaking as the bulletin was being broadcast, this was not done, and any return to the story for updating was given a fresh item number. In both these circumstances a note was made in the last 'comments' column of the logging sheet.

The headline and subheading column were in no sense designed to make possible systematic content notes. Again they were to reflect journalistic practice to be used as *catch lines*. The headline was a basic indicator of the story while the subheading reflected any developments in the story or some other detail. Thus, the Lesley Whittle kidnapping became 'Lesley' and the subhead would detail developments, e.g. 'Brother offers Ransom'. In packages the subhead column would detail the contents of the package. Thus, 'Car

Industry' might yield 'Layoffs at Jaguar', 'Redundancy talks Vauxhall', '25% drop in Car Sales', in the subhead column.

Overall duration of items did include standardised openings and closings, e.g. 'Good Evening'. The only problem encountered was with pick-up lines coming out of the end of one item, e.g. following a film report the newscaster back-announces the name of the reporter. In such cases the back-announcement was timed as part of the follow-item. Items were timed in minutes and seconds, and subsequently were decimalised for computer processing. Hookers (as at the end of the first part of *News at Ten*) and headlines were timed but not, as explained above, given an item number or categorised or included in the total bulletin duration.

The next three columns on the sheet related to television personnel. The project distinguished, as do newsrooms, between newscasters, correspondents and reporters and the numbers entered into these columns were taken from the boxes running across the top of the sheet. Essentially newscasters never leave the studio. If they do they become correspondents or reporters. In ITN during the period of the study, there were three reporter/newscasters and they were entered onto the sheet according to the function they were performing in the particular bulletin. Correspondents were normally introduced as such. If a reporter was introduced as a correspondent he was noted as a correspondent for that bulletin, not as a reporter. Not all personnel are named. If they were not named, 'un-named' was entered into the reporter box, one entry for every un-named person. Naming could be either by the newscaster or correspondent mentioning the name in script or by the use of a caption giving the name (a supercaption). The reporter/correspondent division does not always reflect the hierarchy within the organisations but does indicate slightly different functions on the screen. Correspondents, especially the Westminster and economic staff are often reported by the newscaster as explaining the significance of information they do not themselves appear reporting. Reporters, unless eyewitnesses, are never so quoted. In other words the correspondents are partially thus accorded the status of outside experts. At its most extreme they were often interviewed by the newscasters on BBC2's *News Extra*. BBC regional reporters, reporters belonging to other commercial television companies, when named were entered, as were reporters belonging to foreign networks. Residual categories were created for the purposes of computing; un-named foreign reporters for members of other networks; un-named personnel for other ITV reporters;

'sports commentators', for those working for ITV network sports programmes. The newsroom's own sports correspondents were treated like other correspondents.

The three columns 'IN', 'V/O', 'Int' stand for 'in-vision', 'voice-over' and 'interview' respectively. It was decided to note the appearance of personnel in this way because although presentation seeks to establish the personnel as personalities, it was also known that one criterion of effectiveness used by the broadcasters was the extent to which the so-called 'talking heads' 'in vision' could be avoided in the bulletin. 'In vision' means seeing the newscaster, correspondent or reporter talking to camera whether in the studio or outside on film or videotape. 'Voice over' means cutting away from that talking head while the voice continues over film, videotape, photograph or graphic for however short a time, or indeed never seeing the face at all. 'Interview' means that the personnel concerned conducted an interview whether they were heard or seen during the interview or not. If they were seen asking a question or nodding – the standard device for editing answers to avoid unsightly jumps – this did not count as 'in vision'. In each of these columns the number of the box at the top of the sheet was entered as appropriate.

Interviewees were deemed to include any person not employed by the broadcasting institution who was allowed to speak on camera. This, therefore, included press conferences, speeches, overheard arguments. If no name was given, 'un-named' was entered in the 'name' column. The 'status' column was for indicating why the person was on the news. Thus, if Lord Wigg was being interviewed on the Common Market Campaign it was his status in that campaign not his connection with the Betting Levy Board that was noted. A short categorisation list was generated for interviewees and it was placed at the top of the sheet as an aide-mémoire.

1 Central figure. This applied basically to human interest stories or to situations such as the arrival in London of President Makarios following the attempted coup in Cyprus.
2 Spokesman. This applied to all figures not so directly involved; all officials, police officers, public relations and press officers, politicians. The vast majority of news interviews are therefore with spokesmen.
3 Opposition spokesman. It was only easy to determine an opposition spokesman in two circumstances; first, where such an interview directly followed a 'spokesman' interview, for example, as when management follow union or vice versa; second, where a

member of the parliamentary opposition appeared criticising government policy. This would place him or her in the opposition category whether a government spokesman appeared or not. However in practice it became difficult to operate the distinction between spokesman and opposition spokesman in any other circumstances.

4 Vox pop (vox populi – as used by the broadcasting industry). Classically this is a series of short remarks from people in the street on a given topic but we used it to identify all situations where more than one person was interviewed, without being more than generally identified; most typically this would include people in the street but also groups of workers either filmed collectively – any short sequences of separate talking heads where no one is named.

5 Witness. This category includes witnesses to a situation most usually a crime, disaster or act of terrorism.

6 Conferences. Normally this is a situation where more than one reporter is present asking questions and therefore it is normally an occasion arranged by the interviewee for the purposes of passing on information. It was also deemed to include situations in which speeches are made but not situations where the camera crew overhears private conversation. Thus, all speeches at party conferences, although these are not designed as press conferences, were included in this category. Under this head all the other categories in this list could also be included. The category 6 never stood by itself.

7 Expert. Initially it was felt that the expert was more the province of the current affairs programme but occasionally the news turns to experts for information. During the period of our sample BBC2 *News Extra* frequently treated its own correspondents in this way as has been noted, and *First Report* on ITV also made use of outside experts, normally in the field of science or foreign affairs and normally for the purposes of offering explanations of the events reported on in the bulletin. Experts could normally be distinguished from spokesmen by the way they were introduced, the word 'expert' often being used in the script or implied in the supercaption.

The remaining columns, headed Technics, dealt with a whole range of technical inputs available to the newsrooms. The first four columns dealt with film. Most inputs from locations other than studios will be on film.

During the period of the study the newsrooms were beginning to grapple with a revolution in technique which involved the greater utilisation of videotape from Outside Broadcast units. Until very recently the use of Outside Broadcast units required three lorries, a crew of more than twenty men and many hours of preparation to bring the unit on line. It therefore represented major utilisation of resources and was typically only used for important breaking stories. Now much smaller and more malleable Outside Broadcast units are coming into use but for the period of the study it is true that film was generally still quicker and, with normal two-man units, less demanding of resources than Outside Broadcasts. The great advantage of the Outside Broadcast (OB) is that, when rigged in, it can be used to give instantaneous coverage from an outside source.

Therefore, although this change (to Electronic News Gathering – ENG) was taking place, in filling out this part of the basic logging sheet we assumed film was used unless the story itself revealed that it was either a live input or that it had come in less than one hour before the bulletin. In the latter case it had to be an OB because the process of developing and editing film takes normally longer than one hour.

Even without overt clues, with practice it was felt possible for the research team to distinguish film from electronic sources since the quality of the picture is obviously different. But, there again, during the period of the study this was becoming more difficult since new film stocks had been introduced with the specific purpose of better matching the quality of film to electronic output.

'SOF', 'F/X', 'BW' and 'Pool', mean 'Sound of film', 'Effects', 'Black and white' and 'Pool'.

SOF. In technical terms almost all film now used on television is SOF; that is to say it is recorded by light-weight silent running cameras ('self blimped') operating in synchronisation with light-weight battery recorders, a system perfected in the early 1960s and now widely used. The newsrooms also use a previously perfected system whereby both sound and picture are recorded in the camera by means of a recording head onto a magnetic stripe running down one side of the film. Whether single system (the magnetic stripe) or double headed (the separate tape recorder) both systems allow all the sounds of a location to be recorded. The newsrooms still use the single system because, although it only allows for crude editing and can be spotted because of this, it nevertheless gives the possibility of greater speed since with the double-headed system the sound has to be transferred from ordinary $\frac{1}{4}$ inch magnetic tape to special 16 mm magnetic film before editing can begin.

We were not interested in the above distinction and therefore slightly misused the technical expression 'SOF'. By it we indicated speech which the viewer was allowed to hear uninterrupted by other noises. Thus all interviews were 'SOF' as were all reporter or correspondent pieces to camera, speeches, slanging matches etc. But if at any time commentary drowns out the speech then the film is not, for our purposes, 'SOF'.

F/X. Here again a technical abbreviation was being consciously misused by the project. Before the introduction of the filming systems mentioned above, F/X meant all non-speech noises added to pieces of silent film. We used it to indicate film with commentary spoken over whether it had noises of its own or not. This included speech when overlaid by commentary.

'Black and white'. Black and white film only appeared when it was taken from the library of newsfilm which both broadcasting institutions maintain or when it came from a source which had not yet colourised. These included the most remote studios in the network (Aberdeen) as well as some foreign sources.

'Pool'. We knew there to be an elaborate system of obtaining film from outside sources primarily the European Broadcasting Union (EBU). Each day newsfilm is passed from the participating European national networks to all other EBU affiliates. Whether the film in question was 'pool' or not could only be determined by comparing BBC with ITN. If the film on all channels was the same it came from a common source, such as the EBU news exchange. Such film is passed round the EBU network electronically and not physically. It is therefore recorded onto videotape by the accepting network and transmitted in that form. For our purposes, however, we noted it in these columns rather than in the subsequent videotape columns because in its original form it was film and not tape. In the table this is 'Pool A'.

The next eight columns refer to possible electronic inputs. This did not include noting the basic mode of the bulletins which were known always to be live and from a central news studio. The eight columns therefore refer to inputs analogous to film inputs. If they are electronic they can themselves be live or stored on tape (VTR). Normally live inserts come from other correspondents and reporters in the same studio. Clues as to whether it is live or VTR can be seen at the beginning and end of an insert – newcaster looking off camera to monitor, obvious reactions to the cue to start talking on the part of the correspondent/reporter are the most evident. If the insert begins with an interviewee the clue is yet more apparent than this. If the interviewee talks before the interviewer then it must be VTR since no amateur performer, interviewee, would be trusted to take the cue to begin. The cue must be given to the interviewer and if it is not seen it means that the interview must be recorded so that the front of the interview could be lost and the VTR started on the answer. Other clues as to the use of tape are that if an identical studio interview appears throughout the day although there may be some ambiguity as to the mode of its initial appearance thereafter it must be VTR. And this also applies to reports from correspondents and reporters. If the report does not change from bulletin to bulletin then clearly each use subsequent to the first must be VTR.

In the event it was found that this was not as meaningful a distinction as had been anticipated. The assumption was that the use of VTR represented an increased importance in the minds of the newsroom. But our observational study subsequently showed that the use of tape is so well integrated into the routine practices of the studio as to make the assumption unworkable. The problems involved in distinguishing 'live' from 'VTR' were eventually seen as being of minor importance.

The next six columns can be either 'live' or 'VTR' although the tendency is for the last three, 'EBU', 'Sat' and 'Pool', to be on tape.

'St'=Studio. This involves noting the use of any other television studio in the United Kingdom. It is normally introduced as being from another studio and often the backing in that studio includes a direct reference to where it is, e.g. 'BBC Leeds' on the flat behind the reporter's head. For the BBC this means all BBC regional studios and for ITN the studios of the fifteen commercial television companies. It also means those studios in Westminster and Broadcasting House which are operated by the BBC remotely from their central news studios at the Television Centre in west London. It also means studios established for particular purposes as at the party conferences each year.

'REM'=Remote. Like SOF and F/X this is a technical term which we use unconventionally.

In American television parlance Remote means the same as Outside Broadcast does in the UK. It is not used in the UK at all. 'Remote' in our usage means a television studio either permanent or temporary in any other country, normally within the EBU or in the USA. It is the foreign equivalent of 'Studio'.

'OB'=Outside Broadcast Unit. A mobile television unit carrying with it camera equipment, lights, sound equipment and videotape recorder, requiring some twenty odd men to operate and many hours to bring on line. It can be used either in or out of doors and it sends signals back to the studio by microwave link. Its output can be used live, recorded at the central studio or recorded at the OB. The new generation of OBs mentioned above (ENG) represent a major breakthrough in television technology since the equipment has been made very much lighter and more mobile, requiring a van instead of trucks and five men instead of twenty. During the period of the study this equipment was being experimented with. Normally the Outside Broadcast units use is indicated in the introduction, either by direct reference to it or by the announcement that we are being taken to a remote point for a live report. It represents a major decision as to the importance of a story to commit the OB to it, although as the new generation of equipment is introduced this will become less true. Eventually, some argue, such Electronic News Gathering (ENG) will replace film in the newsrooms altogether. Apart from direct references of this sort, OBs can be spotted because of the quality of the picture which better matches that of the studio than film and because they have more than one camera at a location, cuts can be made on continuous action; e.g. an interviewer interrupting an interviewee, where we cut to the interviewer as soon as he starts to talk. This cannot be done with a single film camera.

EBU=European Broadcasting Union (or Eurovision). The European television networks exchange news stories on a daily basis. These are usually film in the first instance although they may be transmitted after transference to tape. In receiving countries they are always recorded onto tape since the exchange process involves utilising the cable and microwave linking system of the EBU. Eurovision joins its eastern European equivalent Intervision in Prague. If pictures originate in Eastern Europe, which includes the whole of the USSR to the Pacific, then it was noted in this column and a further note 'Intervision' was made in the 'comments' column. The clues as to the use of the Eurovision news exchange material are, first, that the same footage will be used on BBC and ITN, although often, if there is enough material, two services may transmit it in slightly different ways. Second, there will be a voice-over commentary from the studios, since news change material is passed without commentary. Third, the logistics of the material might well indicate that it could only have arrived via Eurovision; that is to say, film of events in a distant part of Europe which could not have reached London and been processed as film in time for the bulletins. Last, it could well be black-and-white since some EBU services are not yet colourised. If the REM column has been ticked it is certain that if the reporter or correspondent is in a studio in Europe, the picture must have come via the EBU link. This involved a tick in the EBU column as well since the factor being highlighted in all these technic columns is expenditure. The news exchange which happens on a 'multilateral' basis is cheaper than a 'unilateral' because the fixed cost of lines, etc. are shared by all participating networks in the former instance but not the latter. This unilateral use of the system is a clear indicator as to the importance of a foreign story in the eyes of the newsroom.

SAT=Satellite. Once all pictures arriving by satellite were announced as such. Now with the firm establishment of the system it is only when the picture quality degenerates that the 'via satellite' caption is used. There is no daily exchange of material via the satellite system analogous to the EBU system. It is extremely expensive even when used multilaterally. Opening a satellite for multilateral use or, even more so, for unilateral use, represents considerable expenditure and is therefore used sparingly and this suggests we may regard its presence as an indicator of importance. Even if not announced its use can be spotted by the same criteria as with EBU above – use of commentary, shared pictures, black-and-white pictures and the impossibility of obtaining pictures from the far corners of the globe quickly by any other means. Most of the world's major sources of stories are within reach of satellite stations, sometimes even when there is no really effective parallel print news agency. The cost of the satellites make them prohibitively expensive to use so that the physical transportation of film back to base, if it is consistent with the importance of the story and the speed deemed to be necessary, will be preferred. Satellite and EBU are modes of communication and can therefore be ticked in conjunction with a number of other columns, most notably film where either system is used to transmit material originated on film as described above.

'Pool'. This indicated the use of the EBU or satellite for some joint (multilateral) coverage which itself originated on tape. The most obvious examples being an American televised press conference or rocket launches. Other pooled material originating as film should be noted in the Film/Pool column. In the tables this is 'Pool B'.

The next three columns represent inputs requiring lower expenditure than the others thus far mentioned in the technics section. They are 'Lib', Library, 'RC', radio circuit, and 'GR', Graphics.

Lib. = Library. Both institutions maintain extensive archives of film material and in this column the use of such previous filmed material was noted. However, determining whether it came from the library or not required using the following guidelines. Library material was easiest to spot when the film was referred to in the script or by a dated caption as not being current. Occasionally it could be spotted because it was cut into current film but was black-and-white whereas the rest of the story was colour. In oil rig stories from Aberdeen the reverse was true (i.e. the library material was colour). In obituaries, film of the deceased must obviously be library.

More difficult to determine were the use of stock shots. The oil rig is but one example. Stock shots would include general footage of neutral material, i.e. general shots of building sites or car assembly lines and the like. Normally such shots would be used in general stories which did not have other current film in them. Noting the presence of library film in this column also meant noting its form in the film columns; i.e. B/W or SOF, etc.

Although the above concentrates on film library there is also an archive of videotape which includes important political speeches or interviews that are deemed to have life beyond the immediate. These are a rare input. If used, the appropriate VTR columns were ticked as well as this library column.

RC = Radio circuit. This is the cheapest way of using foreign correspondents. It is therefore far commoner than REM. By using the radio circuit or at a pinch, an ordinary telephone line, the reporter can broadcast voice only into the programme live or onto audio tape. This is normally illustrated by the use of captions. If Radio circuits are used to get the most up-to-the-minute commentary onto film (which can sometimes be spotted because of the quality of voice reproduction involved), that was not noted. It was registered as V/O and F/X in the film columns. Occasionally for a breaking story which happens too late or is not important enough for any use of the OB, the reporter will phone into the programme live from within the UK. This also happens if a story breaks in areas where there are no television facilities. In each of these cases RC was noted.

GR = Graphics. This included all non-moving images put onto the screen. Eight categories were distinguished and an aide-mémoire printed at the top of the sheet. The categories are as follows:

1 Cap Sup = Supercaption. The 'nameplate'. It was noted for interviewees and personnel. It also includes any superimposed writing placed over another film, tape or live image.
2 Pix = Photographs. These are of two main sorts: portrait shots of news personalities and general news photographs. The differences between them was not noted after the initial drafts of the logging sheet. There are old news pictures which are used in much the same way as library material. The distinction here was also ignored.
3 Graphic. Work of the graphics department beyond the 'nameplate'. This includes all graphs, cartoons, logos, and writing that is transmitted without being superimposed onto another image, e.g. football results.
4 Anim Gr. = Animated graphic. The greatest amount of effort by the graphic department. It is in essence the same as category 3 except that it moves; normally by pulling away strips of the top layer of the work to reveal further information beneath. This can be used to enliven such matters as the Budget.
5 Map. This is a self-explanatory category, although it should be noted that if any movement is involved (i.e. to illustrate the progress of a battle) then it was noted as graphics category 4 not 5.
6 Wire = news photographs received telegraphically. Much of the criteria for spotting these was that used in the case of EBU and Satellite. They are black-and-white, often of poor definition and their provenance indicates they could not be received by any other means. It is a relatively expensive way of getting photos to the screen. We assumed that unless there were internal clues most European and all British photographs would be category 2 not 6.
7 Cap Seq. = Caption Sequence. This means a set of photographs and graphics used as film (action stills). These often require considerable rehearsal time in the studio. This category does not mean a simple sequence of portrait photographs used one after another to illustrate, say, a political argument. It involves camera movement relative to the graphic material and will normally last for more than thirty seconds. At the outset of the project we expected to find caption sequences entirely limited to illustrating scientific stories.

RC Cap = Radio Circuit Caption. Both ITN and the BBC have a particular sort of graphic to cover voice only reports sent by telephone or radio circuit. This normally consists of a small photograph of the correspondent or reporter concerned imposed on some iconic photograph of the place he is speaking from with his name and the place written below.

The next two columns were for the internal use of the project.

T/F = Transfer. A tick in this column indicated that the logger felt the item should be transferred to the archive. The process whereby such decisions were checked, is noted in Chapter 2.

Archive. In this column the technicians noted onto which video-cassette the item had been transferred. The video-cassettes were numbered sequentially and each item was separated on transfer by a further number. Thus 263 ITN 2200 item No. 6 (the 6th item in *News at Ten* on 26 March) would become AVC 35 No. 15. The technicians used a simple videcon camera and handwritten cards to put the archive item numbers onto the cassettes.

The last column was for comments. Many of the suggested areas that might be noted have already been described. In addition we noted here the use of women and black personnel, out of the ordinary caption descriptions of interviewees, material from unusual sources such as amateur film; and made general comments as to exceptionally lengthy or complex items, poor quality of image, mistakes in presentation and so on.

1 Reviewing the News

1. James Curran, 'The Impact of TV on the Audience for National Newspapers, 1945–68' in J. Tunstall, *Media Sociology*, Constable, London, 1970.
2. T. Pateman – Unpublished manuscript, 'Television News: Contexts and Modes of Emission and Reception'. Cf. also his useful analysis and assessment of 'Television and the February '74 General Election', BFI, London, 1974.
3. C. J. Goodhart, A. S. C. Ehrenberg, M. A. Collins, *The Television Audience*, Saxon House, Farnborough, 1975, p. 50.
4. They developed the 'duplication of viewing law' to explain patterns of viewing for all kinds of programme. This states that 'The proportion of the audience of any television programme who watch another programme on another day of the same week is directly proportional to the rating of the latter programme (i.e. equal to it times a certain constant)', *op. cit.*, p. 11.
5. Source: *ITV 75: Guide to Independent Television*, London, 1975, p. 111.
6. *Annual Review of BBC Audience Research Findings*, no. 1, 1973–4, BBC, London, p. 42.
7. BBC Document VR/73/416, London, July 1973.
8. Including Goodhart et al., *op. cit.*
9. BBC document, *op. cit.*, p. 16.
10. A. Smith, *Shadow in the Cave*, Allen & Unwin, London, 1973, p. 109.
11. *Ibid.*, p. 109.
12. ACTT Television Commission, *One Week*, ACTT, London, 1971, p. 11.
13. *Ibid.*
14. Smith, *op. cit.*, p. 110.
15. R. S. Frank, *Message Dimensions of Television News*, Lexington Books, Indianapolis, 1973, p. 11.
16. *Ibid.*, p. 64.
17. *Ibid.*, p. 64.
18. Cf. J. T. Klapper, *The Effects of Mass Communication*, Free Press, New York, 1960.

19 Cf. *The Manufacture of News*, ed. S. Cohen and J. Young, Constable, London, 1973; and Blumler, McQuail and Brown, *The Television Audience*, in *Sociology of Mass Communications*, ed. D. McQuail, Penguin, Harmondsworth, 1975.
20 D. W. Smythe, 'Agenda Setting: The Role of Mass Media and Popular Culture in Defining Development', *Journal of the Centre for Advanced Television Studies*, 3 February 1975, p. 34.
21 Gaye Tuchman, 'Objectivity as Strategic Ritual: an examination of Newsmen's Notions of Objectivity', *AJS*, vol. 77, no. 4, p. 666.
22 For fuller discussion of the epistemological problems here cf. A. Gouldner, *For Sociology*, Allen Lane, London, 1973, chs 1 and 2.
23 K. Nordenstreng, 'Policy for News Transmission', *Educational Broadcasting Review*, August 1971; reprinted in McQuail, *op. cit.*, p. 390.
24 Erving Goffman, *Frame Analysis*, Harper, New York, 1974, p. 14.
25 Stuart Hall, *The Determinations of News Photographs*, Working Papers in Cultural Studies No. 3, University of Birmingham, Autumn 1973, p. 77.
26 Lazarsfeld and Merton, 'Mass Communication, Popular Taste and Organised Social Action', in *Mass Communications*, ed. N. Schramm, 1960, p. 505. University of Illinois Press, Urbana.
27 S. Hood, 'Politics of Television', in McQuail, *op. cit.*, p. 418.
28 *Ibid.*
29 T. Burns, 'Commitment and Career in the BBC', in Sociological Review Monograph No. 13, ed. P. Halmos, Keele, 1969, p. 37.
30 Asa Briggs, *History of Broadcasting in the U.K.*, vol. 1, Oxford University Press, London, 1961, p. 365.
31 *Ibid.*, p. 366.
32 *Ibid.*, p. 374.
33 N. Garnham, *Structures of Television*, British Film Institute, London, 1973, pp. 25f.
34 A. Piepe, M. Emerson, J. Lannon, *TV and the Working Class*, Saxon House, Farnborough, 1975.
35 H. M. Enzensberger, *The Consciousness Industry*, Seabury Press, New York, 1974.
36 R. Miliband, *State in Modern Society*, Weidenfeld & Nicolson, London, 1969, p. 238.
37 G. Gerbner, 'Mass Media and Human Communication Theory', in McQuail, *op. cit.*, p. 51.
38 J. Halloran, P. Elliot, G. Murdock, *Demonstrations and Communication*, Penguin, Harmondsworth, 1970, p. 309.
39 See *BBC Audience Research in the United Kingdom: Methods and Services*, BBC, London, 1970. Robert Silvey, the former head of audience research at the BBC, has given an account of his experiences in *Who's Listening?*, Allen & Unwin, London, 1974. Audience measurement for ITV is provided by an independent research organisation, Audits of Great Britain Ltd, through the Joint Industry Committee for Television Advertising Research which is responsible for the service.
40 S. Hall, 'World at one with itself', in Cohen and Young, *op. cit.*

41 J. Galtung and M. Ruge, *Structuring and Selecting News* in Cohen and Young., *op. cit.*
42 Halloran et al., *op. cit.*, p. 302.
43 Association of Chambers of Commerce Evidence to the Committee on the Future of Broadcasting, reported in *The Times*.
44 Cf. A. Gamble and P. Walton, *Capitalism in Crisis*, Macmillan, London, 1975, Ch. 1.
45 See papers in *Directions in Sociolinguistics*, ed. J. J. Gumperz and D. Hymes, Holt, Rinehart & Winston, New York, 1962.
46 See for example R. L. Birdwhistell, *Kinesics and Context: Essays on Body Motion Communication*, University of Pennsylvania Press, Philadelphia, 1970.
47 Geoffrey N. Leech, *English in Advertising: A Linguistic Study of Advertising in Great Britain*, London, 1966, p. 7.
48 J. M. Sinclair and R. M. Coulthard, *Towards an Analysis of Discourse*, Oxford University Press, London, 1975.
49 John Ellis, 'Made in Ealing', *Screen*, Spring 1975, p. 79.
50 Stephen Heath, 'Film and System', *Screen*, Summer 1975, p. 113.
51 Heath's article 'Film and System', *Screen*, Spring 1975 and Summer 1975 *passim* is an example of these failings.
52 R. A. Pride and G. L. Walmsey, 'Symbolic Analysis of Network Coverage of the Laos Invasion', *Journalism Quarterly*, vol. 49, Winter 1972, pp. 539–43; K. S. Thompson, A. C. Clark, S. Dinitz, 'Reactions to My-Lai: A Visual-Verbal Comparison', *Sociology and Social Research*, vol. 58, no. 2, January 1974, pp. 122–9.
53 E.g. BBC1 and ITN main bulletins, 24 March 1975; see pp. 240–1 above.
54 Frank, *op. cit.*, p. 28.

2 Constructing the Project

1 ACTT Television Commission, *One Week*, ACTT, London, 1971.
2 Cf. P. Hartmann, *Industrial Relations Journal*, vol. 6, no. 4, Winter 1975, pp. 4ff. for a contrary view.
3 F. Rositi et al., *News and Current Events on TV – an Analysis of Television News on Four European Television Organizations*, Prix Italia Report, Florence, 1975.
4 Statistical Package for the Social Sciences (SPSS) is fully described in the *SPSS Manual* by N. Nie, D. H. Bent and C. H. Hull, McGraw-Hill, New York, 1970. The project made use of Version 5 of SPSS on the IBM 370 computers in Edinburgh and Newcastle.
5 Glasgow Media Project, *Evidence Presented to the Committee on the Future of Broadcasting: Television Coverage of Industrial Relations*, July 1975.
6 There was one interview after the day they returned to work.
7 'Television Coverage of Industrial Relations: A BBC Analysis of the Glasgow University Media Project Report', December 1975.
8 'A Response to "A BBC Analysis of the Glasgow University Media Project Report" ', February 1976.

9 Sir Charles Curran, *Broadcaster/Researcher Co-operation: Problems and Possibilities*, International Seminar Report, ed. J. D. Halloran and M. Gurevitely, Centre for Mass Communication Research, Leicester, 1971, p. 48.

3 Inside the Television Newsroom

1 S. Hood in D. McQuail, ed., *Sociology of Mass Communications*, Penguin, Harmondsworth, 1972.
2 Galtung and Ruge in S. Cohen and J. Young, eds, *The Manufacture of News*, Constable, London, 1973.
3 10 feet by 12 feet.
4 All BBC buildings having a bar facility operated as a part of the 'BBC Club', subscriptions for which are automatically deducted from salary.
5 E.g. EBU Exchange stories rise from an all-bulletin average of 1·6 per cent (BBC1), 1·9 per cent (ITN) and 1·5 per cent (BBC2) of all items to 3·8 per cent, 3·8 per cent, 4·6 per cent respectively at weekends.
6 Result of court decision on subpoenaed film.
7 ITN film does contain a significantly lower proportion of 'talking heads' than does BBC's (see above pp. 116ff).
8 Lord Hill, *Behind the Screen*, Sidgwick & Jackson, London, 1974, p. 207.
9 See below p. 141.
10 *Inside the News* (BBC2, 11 October 1975) recalled that at the outset BBC used staff journalists to present the bulletins with disastrous results. Perhaps this is why the thespian tradition grew up.
11 J. Halloran, P. Elliot, G. Murdock, *Demonstrations and Communication*, Penguin, Harmondsworth, 1970.
12 Hill, *op. cit.*, p. 207.

4 Measure for Measure

1 P. Rock, 'News as Eternal Recurrence' in S. Cohen and J. Young, eds, *The Manufacture of News*, Constable, London, 1973.
2 M. Frayn, 'The Tin Men' in *ibid.*
3 A study of news and current affairs programming in the USA concluded that 'The quantity of news and public affairs programming shown, the types of programming offered, the scheduling factors associated with these programmes, and even the content of the programming were largely explicable in economic terms. . . . The use to which air time was put was basically a function of the quest for profit and only secondarily a function of other factors.' There is a close relationship between audiences and profits. P. Wolf, *Television Programming for News and Public Affairs*, Praeger, New York, 1972, p. 137.
4 For a 'professional' view see, for example, Paul Fox, 'One Hour of Television News?', *The Listener*, BBC, London, 20 November 1975.

5 Taking all weekday bulletins there is a correlation of +0·51 (s= 0·001) between the number of items and the length of the bulletin, i.e. length does not account for the whole of the variation in the number of items.
6 John Birt and Peter Jay have claimed that current television journalism contains a 'bias against understanding' which can be attributed to its antecedents in newspapers and documentary film (*The Times*, 28 February, 30 September, 1 October 1975). They imply that one condition for eliminating the bias would be an extension in the length of the news and recommend a daily, one-hour programme containing news headlines and detailed treatment of about six main stories.
7 This does not mean that interviews normally last longer than 4 minutes but that they will be part of items that do. Three out of ten news items contain interviews.
8 Franco Rositi, *The Television News Programme: Fragmentation and Reconstruction of our Image of Society*, Prix Italia Report, 1975; English version, p. 13.
9 With the exception that the average length of items in the home affairs and science categories on BBC2 arises because many of these items are comparable to feature articles in the press.
10 See for example, Raymond Williams, *Communications*, Penguin, Harmondsworth, 1962, esp. Chapter 3.
11 Cf. Chapter 6 in which it is shown that unions such as ASTMS aim for *First Report*. Given early coverage it is possible that the story will reappear in later bulletins.
12 Scheduled at a very late time, *News Extra* was in the best position to cover parliamentary debates and, in fact, often did so at some length.
13 The standard deviation is calculated by determining the deviations of the scores from the mean for each category, squaring each deviation, finding the arithmetic mean of these numbers and then taking the square root of this mean. The formula for standard deviations is

$$s = \left[\frac{\Sigma_{i=1}^{N}(X_i - \overline{X})^2}{N}\right]^{\frac{1}{2}}$$

where X equals the mean of the original scores.
14 Note the high level of film inputs in the foreign (32) category (see Tables 4.12, 4.13, 4.14).
15 Numerous experiments in social psychology have shown the importance of the 'primacy effect'. This would imply that news which comes first is more likely to be retained, even if subsequent news contradicts the earlier information. For a summary of research into primacy effects, see R. L. Rosnow, 'Whatever happened to the "Law of Primacy"?' in *The Process of Social Influence*, ed T. D. Beisecker and D. W. Parson, Prentice-Hall, New York, 1972.
16 BBC1's rather high proportion (19 per cent) of lead stories in the crime category partly reflects the relatively greater use of the story category on this channel, which is itself a divergence from the three-monthly average arising from the sampling method. This does not contradict the finding that ITN regularly gives more prominence to crime stories.

Even if such distortion could be eliminated, it would not be possible to construct a perfect system of rank-ordering rules from these tables alone because there is an element of indeterminacy which is a product of time and technical constraints as well as the mode of presentation.

17 See Chapter 5.

18 See Appendix 2.

19 The analysis does not distinguish between items with more than one input of a given type and items with only one. It does of course show the range of inputs (cf. Computer note above, Chapter 2).

20 See Appendix 2.

21 The use of radio links (RC) and RC captions provides a simple internal check on the reliability of coding. Variability in the coding of context categories was minimised by the collective checks described in Chapter 2. The coding of technical inputs presented less of a problem, since the categories are precisely defined and are mutually exclusive. The use of RC captions is always concomitant with the use of radio links, although the reverse is not true, since radio circuits may be used with still photographs, graphics, or film. The inputs of RC should therefore be equal to, or greater than, the inputs of RC captions. Inspection of Tables 4.12–4.17 shows this to be true for all categories on ITN. On BBC1 and BBC2 the relationship does not hold in two categories – Northern Ireland and Parliamentary respectively. This can best be explained by the use in political stories of captions superimposed on still photographs. In a few cases these were incorrectly coded as RC captions without noting the use of a radio circuit, since the commentary was recorded by the normal method.

22 See above, pp. 18–31 for a discussion of 'events' and 'facts'.

23 Cf. J. Whale, *The Half-Shut Eye*, Macmillan, London, 1969.

24 A comparison of the figures for each month (not reproduced here) shows that the pattern of use of interviews remains virtually unchanged from month to month.

25 Of these, only 4 were with women workers.

5 Contours of Coverage

1 Chapter 4.

2 R. S. Frank, *Message Dimensions of Television News*, Lexington Books, Indianapolis, 1973, pp. 21–2.

3 *Inside the News* discussion programme transmitted on BBC2; this point was made by both the Editor BBC News and the Editor ITN.

4 E.g. 'The strike at the Chrysler Engine Works at Coventry is to go on. Shop stewards today told the 4,000 workers on strike to go on with their demands for an immediate eight pounds a week rise.' Fourth item in *Newsday* bulletin, BBC2, 14 May 1975.

5 E.g. 'News at home included Chancellor Denis Healey's warning of the lunacy of inflationary wage increases. Unemployment figures were up by 10,000 and there was an even bigger increase in short-time working. Still no end to the big London dock strike but a new container

work offer has been suggested by the government. The rejection by both rail and power workers of pay offers from their boards which are within the Social Contract. Each union is quoting the miners' 30 or more per cent increase as the basis of a claim for similar treatment and with dustcart drivers back in work at Liverpool, but still on strike in Glasgow, the City Corporation called in the professionals.' (There followed a short report with film and interviews on the situation in Glasgow.) *News Review*, BBC2, 23 March 1975.

6 E.g. Newscaster: 'On the day that it has been announced by the Government that new car sales last year were down by 25 per cent on 1973, the Director of British Leyland's Cowley Plant has warned of a calamity if the strike situation there gets worse. Figures out today show that private car and van registrations dropped from 1,688,000 in 1973 to 1,273,000 last year and all vehicle registrations were down nearly as much by 20 per cent. The warning came in a letter from the Plant Director, Mr John Symons, to Leyland employees as the Company and the Engineering Union agreed to talks tomorrow at the Conciliation and Arbitration Service to try to solve the strike of engine tuners at Cowley. Mr Symons said that the strike had meant that Cowley was failing to meet what he called its survival budget. He also gave a warning that a further deterioration would be calamitous with the strongest likelihood of a major reduction in manufacturing and employment at Cowley. Here is our Industrial Correspondent, Giles Smith:'

Smith: 'Mr Symons' warning to Cowley's 12,000 workforce is easily the strongest since this dispute started a month ago. In using words like "calamity" and a "major reduction in manufacturing and in employment" he has got the full authority of Lord Stokes and the Board. The message is simple: if this dispute isn't sorted out very quickly indeed, the government may well choose not to help the company in this difficult time. Meanwhile the problems of the industry as a whole have been highlighted by today's figures [. . .] in sales last year down 25 per cent. This has seen tens of thousands of the industry's half a million work force going onto short time and unsold cars piling up in the car parks. All this in the year when Aston Martin went broke, British Leyland went to the government for help and the three Americans, Ford, Vauxhall and Chrysler, warning they might have to transfer operations to the Continent. Chrysler is probably the worst hit with their Ryton plant near Coventry being the main casualty. Between Ryton and their Scottish plant at Linwood, Chrysler have 10,000 men on a 2-day week. Linwood, in fact, closes next week. 1,000 have lost their jobs altogether, there is short-time working at the Stoke engine plant and they've got 28,000 unsold cars waiting to go to the dealers. Vauxhall, the General Motors subsidiary which should be getting 15 per cent of the market but isn't, is not much better off. At their plants at Luton, Dunstable and Ellesmere Port 6,000 of their workers will be going on 3-day working by the end of this week. Of the Americans, Ford is the least affected, at Halewood and their other big plant at Dagenham, the company sales of Capris and Cortinas continue to go well. There is no short-time working yet though not much overtime either. Even they, however, have laid off 12,000 white-collar workers. The company's main bright spot is the launch tomorrow of their brand new version of the Escort. A car, they hope, will help them weather the energy crisis. The problems of our biggest and only British owned company, Leyland, are well known, particularly at Cowley. Here the jobs of 12,000 are tonight in the balance, and Leyland are not just talking about temporary layoffs until the tuners go back to work.

Elsewhere the company's 20,000 employees at Longbridge haven't worked any overtime for months and Jaguar the export leader have put 7,000 workers on a 4-day week.

All this, the most depressing scenario for an industry whose exports last year totalled nearly £2,000m, nearly one fifth of the country's total overseas sales.

If Britain is to make any sort of attempt at clearing the present staggering Balance of Payments problem, it can certainly not do without a healthy motor industry for the investment, the manpower and the will to defy the energy crisis and step up sales.'
First item, ITN, *News at Ten*, 22 January 1975.

7 We should make clear that in this context we are not measuring the amount of time allocated to these stories. Time allocation of news items has, however, been discussed above. The base for all the counting in this chapter is the complete television news output for the 22-week period. There are two possible sources of distortion which are taken into account. First, there is the nine-day period at the end of May during which ITN was off the air because of an industrial dispute. Second, a small number of the bulletins were not recorded. These accounted for only 5·7 per cent of the total output in the first three

months (see above Chapter 2, p. 48) and the number lost in April and May was even fewer.

8 Cf. P. Hartmann, *Industrial Relations Journal*, vol. 6, No. 4, Winter 1975, pp. 4 ff. for an alternative view based on his own identification of 'major themes'.

9 Official statistics, like the news itself, are 'products' which pose their own problems of interpretation. For a discussion of the epistemological arguments surrounding their use, see B. Hindess, *The Use of Official Statistics in Sociology: a Critique of Positivism and Ethnomethodology*, Macmillan, London, 1973. Our use of the Department of Employment's statistics in this chapter has one main purpose: to demonstrate the existence of an independently derived and publicly available alternative description to that of the news. The conceptual model and technical apparatus used in the production of these statistics, though not beyond criticism, does not pose any major theoretical problem.

10 See Accidents, pp. 193–5 above.

11 'The official series of statistics of stoppages of work due to industrial disputes in the United Kingdom relates to disputes connected with terms and conditions of employment. Stoppages involving fewer than 10 workers or lasting less than one day are excluded except where the aggregate of working days lost exceeded 100. Workers involved are those directly involved and indirectly involved (thrown out of work although not parties to the disputes) *at the establishment where the disputes occurred.* The number of working days lost is the aggregate of days lost by workers both directly and indirectly involved (as defined). It follows that the statistics do not reflect repercussions elsewhere, that is, *at establishments other than those at which the disputes occurred.* For example, the statistics exclude persons laid off and working days lost at such establishments through shortages of material caused by the stoppages included in the statistics.'

Notes on the compilation of dispute figures from the Department of Employment statistics

(1) The monthly index of 'stoppages', 'working days lost', and 'workers involved', published in the *Department of Employment Gazette* (HMSO) is cumulative. Thus figures published in February 1975 show disputes for January 1975; those published in March 1975 show total dispute figures for January and February; and those published in April show totals for January, February and March, etc. The separate monthly indices were thus established by simple subtraction.

(2) The only available figures recording stoppages, days lost, etc., was a classification by industrial sector which contains certain sectors otherwise considered separate. E.g. mechanical, instruments, and electrical engineering (sectors 07, 08, 09), are treated as simply 'Engineering' in the dispute index. Similarly, leather goods (14), and clothing and footwear (15), are combined and insurance, banking, finance, business services (24), professional and other services (25) are combined with public administration (27).

12 N.B. ITN was off the air for one week during the period in the study.

13 See Chapter 7 below.

14 Francis Williams, *Dangerous Estate*, Arrow, London, 1959, p. 107.

15

	Circulation 1974	Readership by social grade*			
	m.	AB	C^2	C^1	DE
Daily Mirror	4·6	6	21	42	32
Daily Express	3·2	15	28	31	27
Daily Telegraph	1·4	42	35	14	10
Guardian	0·4	40	36	16	9
The Times	0·4	5	28	13	9

* By Registrar General Class Studies: Source: JICTARS National Readership Survey, July 1973–June 1974.

16 Thus a possible indication of a further dimension of news value in frequency of Press Association tape items was not utilised.

17 The Press Association carried *one* dispute report in this area in January – no reports in February.

18 Plessey, Liverpool – strike in protest of Post Office contract given to a Swedish firm (BBC1, 17.45 and 21.00, 25 February 1975).

19 There was no television coverage of dispute in this sector in February.

20 The press sample shows coverage of these three disputes in February and two others. The PA also covered those three with an additional three strikes.

21 See below Chapter 7.

22 The margins for error on which these figures are based, are noted on p. 111 of the Annual Report.

6 Trades Unions and the Media

1 Cf. the concept of producer initiative in a 'policy vacuum', Blumler, 'Producer Attitudes towards Television Coverage of an Election Campaign' in J. Tunstall, *Media Sociology*, Constable, London, 1970, p. 416. Or see Elliot: 'Programme content was less a manifest consequence of decisions about its substance than a latent consequence of its passage through the production process itself', P. Elliot, *The Making of a Television Series*, Constable, London, 1972, p. 85.

2 D. Lockwood, *The Black-coated Worker*, Allen & Unwin, London, 1958.

3 B. Roberts, R. Loveridge and J. Gennard, *Reluctant Militants*, Heinemann, London, 1972, p. 63.

4 R. Bacon and W. Eltis writing in the *Sunday Times*, 2 November 1975. See also Michael Meacher MP who, in a separate study, calculated the fall in real income after 1972 in *The Times*, 21 January 1974; *Capitalism and Crisis*, A. Gamble and P. Walton, Macmillan, London, 1976.

5 S. Engleman and A. Thomison, *The Industrial Relations Act*, Martin Robertson, London, 1975.

6 J. Westergaard and H. Resler, *Class in a Capitalist Society*, Heinemann. London, 1975, pp. 418–19.
7 Figures from Westergaard and Resler, *op. cit.*, p. 415.
8 Source: *Department of Employment Gazette*, November 1975. See Appendix II for figures on union growth between 1964 and 1974. In addition, a body of theoretical and empirical work has been developed on the differing work conditions, job security perceptions and attitudes of white- and blue-collar workers.
9 See Lockwood *op. cit.* See also in this area: D. Wedderburn, 'The Conditions of Manual and Non-Manual Workers', *Proceedings of the S.S.R.C. Conference on Social Stratification and Industrial Relations* (January 1969) and J. Westergaard.
10 Quoted in S. Harrison, *Poor Men's Guardian*, Lawrence & Wishart, London, 1974.
11 See H. Pelling, *History of British Trade Unionism*, Penguin, Harmondsworth, 1971. Pelling notes, for example, how isolated acts of violence tended to be seen by the established press as an indictment of unionism as such (p. 65).
12 See Pelling, *op. cit.*
13 The white-collar unions, who increased their membership; but they were still on a fairly small scale. For example, the Association of Engineering and Shipbuilding Draughtsmen grew from 19,310 in 1939 to 36,661 in 1945. See Roberts, Loveridge and Gennard, *op. cit.*, p. 79.
14 Cited in T. Lane and K. Roberts, *Strike at Pilkingtons*, Fontana, London, 1971, p. 81.
15 Our own research suggests that this is a particularly important area for any organisation which is seeking publicity, since inclusion in the lunchtime bulletin often encourages coverage throughout the rest of the day.
16 The CBI have in fact published a pamphlet entitled *Don't Be Afraid of the Box* as part of this campaign.
17 But only in part, since opinions as to its usefulness vary widely amongst trade unionists.
18 G. Strauss, 'The White Collar Unions are Different', in *Harvard Business Review*, XXXII September–October 1954, quoted in R. M. Blackburn, *Union Character and Social Class*, Batsford, London, 1967, p. 52.
19 R. M. Blackburn, *op. cit.*, p. 53.
20 J. A. Banks, *Trade Unionism*, Macmillan, London, 1974, p. 6.
21 G. S. Bain, 'The Growth of White Collar Unionism in Britain' in *British Journal of Industrial Relations*, 1966, p. 330.
22 Figures for local authority employment are from Robert Bacon and Walter Eltis in the *Sunday Times*, 2 November 1975, p. 17.
23 Roberts, Loveridge and Gennard, *op. cit.*, p. 322.
24 *C.O.H.S.E. From Phase Three to Halsbury*, published by the Confederation of Health Service Employees, 1975, Foreword.

25 Hindell argues that this 'respectable soft sell' was part of an early recruitment policy on the part of NALGO. See K. Hindell, 'Trade Union Membership', in *Planning*, vol. XXVII, no. 463, PEP, 1962.
26 See Roberts, Loveridge and Gennard, *op. cit.*, p. 322.
27 NALGO's evidence to the Royal Commission on the Press, p. 7.
28 See Chapter 7.
29 Commenting on this, a writer in the NALGO newspaper *Public Service*, recently noted that: 'Can you tell me, for example, why a top executive needs more leave than Cynthia, who has been pounding a typewriter in our office for some twenty years? The gradualist argument will be: since it has always been, then we cannot change things. Increasingly though, it looks as though economic realities will not let us go on with our well organised lives with any change being on a little-by-little basis. These threaten our jobs and living standards. There is a topic for any annual general meeting. . . . And yet, even that discussion still bears a veneer of élitism. There is something different for the top man (and they are men always, not women). In all the discussions which have been going on about possible staff reductions the threat of redundancy is there for the clerks and the typists. But where is the first chief executive threatened, or head of department? They are proclaiming the sad need for sacrifice, but as ever, the élite is indispensable to see the thing through.' *Public Service*, November 1975, p. 6.
30 The bulk of NALGO workers in local authorities earn less than the national average wage for full-time workers. Nearly 70 per cent earn less than £2900 p.a. and 25 per cent earn less than £2040 p.a. Should settlements be on a percentage basis which would maintain differentials or on a sliding scale which would erode them? The pay settlement which was negotiated in 1975 for Local Government Officers was thus the subject of much contention within the organisation. As we have already commented above, NALGO is in the process of transforming from being essentially a management/staff association to being, by self-definition, something much more like a trade union.
31 As one member put it succinctly to us, 'Your shop steward may also be your boss.' NALGO is thus a rapidly growing union with a membership which is potentially divided over differentials in pay and job security. The extent of the divisions which already exist was graphically illustrated to us by one NALGO district official who commented that almost the only issue which had united the 'right' and 'left' at a recent NALGO annual conference was the determination on both sides not to affiliate to the Labour Party.
32 The television offered a rare glimpse of these processes in the edition of *World in Action*, transmitted during the 1975 TUC conference which concentrated on the right/left split in the NALGO delegation.
33 See Chapter 5.
34 Reproduced by courtesy of the Post Office Engineering Union.
35 NALGO's evidence to the Annan Commission on the Future of Broadcasting – page 1 – supplied by courtesy of NALGO.
36 For a more detailed discussion of this coverage, see below.

37 NALGO, *op. cit.*
38 The ABAS and the ACTT are in the process of amalgamating into the Amalgamated Film and Broadcasting Union (AFBU).
39 Blumler, *op. cit.*; Elliot, *op. cit.*
40 At the end of this period the pay negotiations went to arbitration and the NUR began to differentiate itself from the other rail unions. There are thus some grounds *after* April for the NUR to receive a more distinct coverage.

7 Down to Cases

1 Reginald Bosanquet introducing reports of both the Liverpool and Glasgow drivers' disputes. Item 10, *News at Ten*, 21 March 1975.
2 BBC1, 5.15 p.m., 15 March 1975 and BBC1, 17.45, 14 March 1975 respectively.
3 A total of 96 bulletins (ITN 43, BBC2 17 and BBC1 36) were recorded between 11 January and 14 April. A further 6 bulletins from the BBC were obtained from BBC microfilm to fill the gaps in the coverage analysis caused by machine failure, etc. (16 March 1975, 2 BBC2 bulletins and 3 BBC1 bulletins, and 1 BBC1 bulletin on 14 March 1975). However, as ITN microfilm or other transcript material was not available, it was not possible to check those few ITN bulletins that may have been missed. Thus the sample for analysis was 102 bulletins, 40 bulletins on BBC1, 19 on BBC2 and 43 on ITN.
4 On 11 January, in 2 ITN bulletins, 2 BBC1 and 2 on BBC2.
On 12 January, in 1 ITN bulletin, on 13 January in 3 ITN bulletins, 2 BBC1 bulletins, and on 15 January, 1 BBC1 bulletin.
5 In a film report from Glasgow in this bulletin, Michael Buerk tells us that 'Glasgow dustmen [sic] are on strike again'.
6 Dan Duffy was interviewed on 9 April 1975 after a mass meeting decision to end the strike. This was shown on BBC1 Early Evening bulletin and *Nine O'Clock News* and similarly on ITN's *First Report*. When the dustcart drivers returned to work on 14 April, all 3 ITN bulletins contained interviews with one of the drivers, Mr Docherty, and a binman who said of the strikers that they should have returned to work long ago.
7 There are these further references: 'This dispute was showing signs of becoming one of what Mr Wilson has called "manifestly avoidable stoppages of production in the car industry"' (BBC1, *Nine O'Clock News*, 9 January 1975). And for ITN: 'One factory referred to by Mr Wilson, British Leyland's Austin Morris plant at Cowley in Oxford, has shut down for the weekend, because of industrial trouble. Keith Hatfield reports: The Austin and Morris car plant at Cowley is now totally shut down. 12,000 men have been laid off because 250 engine tuners want their jobs regraded. *It's the kind of strike that has contributed significantly to the dire economic difficulties of British Leyland*' (ITN, *News at Ten*, 4 January 1975; our italics).
8 Supra, Chapter 5, pp. 147–8.

INDEX

www.ingramcontent.com/pod-product-compliance
Ingram Content Group UK Ltd.
Pitfield, Milton Keynes, MK11 3LW, UK
UKHW020411010325
455677UK00029B/844